Incentives in
Water Quality Management
France and the Ruhr Area

Research Paper R-24

Incentives in
Water Quality Management
France and the Ruhr Area

Blair T. Bower, Rémi Barré
Jochen Kühner, Clifford S. Russell
with Anne J. Price

RESOURCES FOR THE FUTURE / WASHINGTON, D.C.

RESOURCES FOR THE FUTURE, INC.
1755 Massachusetts Avenue, N.W., Washington, D.C. 20036

Resources for the Future is a nonprofit organization for research and education in the development, conservation, and use of natural resources and the improvement of the quality of the environment. It was established in 1952 with the cooperation of the Ford Foundation. Grants for research are accepted from government and private sources only if they meet the conditions of a policy established by the Board of Directors of Resources for the Future. The policy states that RFF shall be solely responsible for the conduct of the research and free to make the research results available to the public. Part of the work of Resources for the Future is carried out by its resident staff; part is supported by grants to universities and other nonprofit organizations. Unless otherwise stated, interpretations and conclusions in RFF publications are those of the authors; the organization takes responsibility for the selection of significant subjects for study, the competence of the researchers, and their freedom of inquiry.

Research Papers are studies and conference reports published by Resources for the Future from the authors' typescripts. The accuracy of the material is the responsibility of the authors and the material is not given the usual editorial review by RFF. The Research Paper series is intended to provide inexpensive and prompt distribution of research that is likely to have a shorter shelf life or to reach a smaller audience than RFF books.

Library of Congress Catalog Card Number 81-47032

ISBN 0-8018-2661-6

Copyright © 1981 by Resources for the Future, Inc.

Distributed by The Johns Hopkins University Press,
  Baltimore, Maryland 21218

Manufactured in the United States of America

Published March 1981. $14.00.

TABLE OF CONTENTS*

*A more detailed table of contents can be found at the beginning
of parts II and III.

List of Tables

List of Figures

PREFACE

The two studies reported in Parts II and III of this volume were done
as part of a program of research on implementation incentives for environmental quality management under the aegis of the U. S. Environmental
Protection Agency, Charles N. Ehler, Project Officer, with the cooperation
of the U. S. Council on Environmental Quality, Edwin H. Clark, Project
Officer. Additional support for the French study was provided by the
French Ministere de l'Environnement et du Cadre de Vie. Final preparation
of the volume and writing of the introductory part were supported by
Resources for the Future under a grant from the Alfred P. Sloan Foundation.

Effluent charges as an incentive mechanism in water quality management
in the United States were proposed by Allen Kneese of Resources for the
Future in the 1960s. In doing so he drew particularly on the long experience with such charges in the Ruhr area of the Federal Republic of Germany,
and on the system proposed in France, which system was subsequently
introduced in 1970. Descriptions of the effluent charges systems for
these two areas were contained in the RFF book, Managing Water Quality,
by Allen V. Kneese and Blair T. Bower, published in 1968. Subsequent to
that publication, the charges systems in both areas have been discussed
in other reports, covering basically the same subject matter.

Experience over the last decade indicated that effluent charges
constituted only a part of the systems for water quality management in

both France and the Ruhr area of the Federal Republic of Germany. As
noted in Part I of this volume, mixed systems of regulatory and economic
incentives are typical rather than atypical, both in these countries and
elsewhere in Europe, and in the United States. Consequently, it was deemed
important to provide both a more complete description of the two systems
than was available, and some analysis of the role of effluent charges in
those systems. In addition, it was considered important to be specific
about the ways in which the effluent charges to be imposed upon discharges
were actually calculated. The reality was not necessarily consistent with
economic theory or what many economists had proposed over the last decade
or two. Moreover, it is clear that legislation is a necessary and not a
sufficient condition for effective and efficient water quality management;
institutions and the human beings of which they are comprised, are equally
critical factors. Thus, importance was attached to trying to convey as
complete a picture as possible of all the elements of water quality management in the two areas under consideration.

Possessing a better understanding of these two systems, in which
effluent charges had existed as one element of water quality management
for some time, has become increasingly important in the context of
discussions in recent years in the United States of "regulatory reform".
It is one thing to state that regulatory reform is necessary, but quite
another to be able to point to specific experience that would indicate
the direction in which such reform should go if improvements are to be
made in the effectiveness and efficiency of water quality management.

Water resources management, as noted by Lyle Craine in his Water
Management Innovations in England, involves a number of tasks/functions/

activities, which are carried out by various public and private entities.
It is the set of tasks by which the desired end is accomplished, whether
that end be delivery of irrigation water or the achievement of ambient
water quality standards. The physical measures for reducing discharges,
the implementative incentives which induce the public and private entities
to adopt the physical measures, and the institutional arrangements which
allocate the tasks, comprise the necessary elements. It is within this
context that the two studies reported herein were undertaken.

November 1980   Blair T. Bower

ACKNOWLEDGMENTS

This volume would not have been possible without the contributions of a number of individuals. With respect to the French study, sincere thanks are due to Messieurs Teniere-Buchot, Caille, and Vandewoestyne of the Seine-Normandie Basin Agency, and Monsieur Billecocq of the Center for Pollution Control of the Seine-et-Marne Department, for having spent so many hours in discussion of the contents during its preparation. Messieurs Teniere-Buchot, and W. R. D. Sewell of the University of Victoria, Canada, provided detailed and much appreciated reviews of the study in draft form.

With respect to the study of the Genossenschaften, special thanks are due to Messieurs Albrecht and Gaebel, and Dr. Imhoff of the Ruhrverband, Professor Rincke of the University of Darmstadt, and Baudirektor Boucsein of Regierungspräsidium Arnsberg, who discussed the structure and details of the systems and read various versions of the draft report. Numerous people in various industrial activities provided necessary input, but were promised anonymity and therefore cannot be acknowledged by name. Renate d'Argeucelo and Alice Kühner provided help throughout the study.

This volume was produced by Anne Price from a melange of French, German, and American typographical and editorial styles. The final result is a tribute to her fortitude and patience. Pathana Thananart drew the figures.

PART I

INTRODUCTION

Clifford S. Russell and Blair T. Bower

Chapter 1

EFFLUENT CHARGES AND OTHER IMPLEMENTATION

INCENTIVES IN WATER QUALITY MANAGEMENT: AN OVERVIEW

In the continuing debate about government regulatory activity,
a major bone of contention is the relative merit of two sharply con-
trasted methods for achieving public goals through the direction of
private and public activity. In one of these, a governmental agency
at the national or state level tells firms, enterprises, individuals,
and governments of lower jurisdiction just what they must do--what
technology they should install, where they should locate, how they
should design a building or a machine. This is often referred to as
"command and control" regulation.[1] In the other approach to regulation
the government charges the activities of interest a fee per unit of
undesirable behavior (as an emission fee on discharges of pollutants),
or offers a subsidy per unit of desirable behavior (as a subsidy on
recycled material, or per unit of low sulfur fuel burned), or defines
appropriate property rights so that a market can establish a price for
desirable or undesirable behavior (as a marketable permit to discharge
pollutants). These regulatory devices are lumped under the broad
heading, "economic incentive systems."

Economists are wont to stress that these latter systems allow
individual decision makers flexibility in reacting to the incentives

offered, and that this at least improves the chance of achieving the goals of the regulation at least cost. Further, the economic incentive systems are said to provide a spur to technological advance in the desired direction (e.g., cheaper, more effective means of reducing discharges); and in some configurations these systems can reduce the amount of information that must be available to government regulators. Partisans of the command and control variety of regulation, on the other hand, stress that by using suitable commands the government can achieve with certainty its regulatory goals, because it will not be subject to the possible ignorance or perverseness of the regulated activities, qualities that can destroy the effectiveness of the economic incentives.

A first difficulty with this debate has been and continues to be that the opposing sides do not carefully specify all the relevant features of the systems they attack and defend. And the details make a very great deal of difference in what can be expected by way of performance.[2] A second and more fundamental problem is that the debate focuses attention almost entirely on only a subset of the range of components that must be put together to produce a complete system of implementation incentives. Finally, the debate over command and control versus economic incentives suffers from being conducted in a near perfect vacuum of experience with actual use of the most frequently mentioned economic incentive systems.

This book is addressed to the second and third of these problems. In this introductory chapter, an attempt is made to broaden the horizon of debate by describing all the components of an implementation incentive system, showing how they fit together, and suggesting just how wide a range of choice is really available to would-be system designers. An implication of this discussion is that it will be a rare implementation incentive system that does not display a mix of command and control and economic incentive features. The remaining two major parts of the book contain detailed discussions of the operation of two such systems: the French national system for water quality management, and the German regional system used in the river basins of Nordrhein-Westfalen. These systems were chosen for such careful attention because they are frequently cited as examples of effluent charge systems, and in that capacity have been of great potential interest to advocates and critics of this incentive device.[3] The discussion in these latter parts will, however, make clear two points. First, the characterization, "effluent charge system," does not begin to capture the richness and complexity of either of these two systems. Second, the reader will find that the effluent charge part of these two systems has little in common with any of the several varieties of this device most often advocated for the United States. Thus, if the book succeeds in its purpose it will make life harder, not easier, for participants in the debate over how to implement environmental quality goals. It will demonstrate both that to argue economic incentives as against command and control is seriously to oversimplify; and that the oft-cited French and German experience with effluent charges is really experience with complex mixed systems, including cost-based fees for service, but not effluent charges in the sense that label is commonly used. The vacuum of relevant empirical evidence therefore remains a serious difficulty.[4]

IMPLEMENTATION INCENTIVE
SYSTEMS

There is more to achieving the goals of a piece of pollution control legislation than announcing either an effluent charge of E cents per pound of discharge, or specifying which types of treatment equipment will be installed by which sources. Even if the choice of the basic policy instrument were confined to those two alternatives, however, there would remain two other points to worry about. One is how to monitor the performance of the activities where behavior is the issue. The second is how to encourage compliance with whichever system of instruments and monitoring rules has been adopted. The potential richness of available implementation incentive systems arises in part from the variety of possibilities for dealing with these two concerns: monitoring and sanctions.

But the fundamental choice is actually never so narrow as that posed above. Rather, there is a wide range of instruments through which a government can seek to direct or influence the actions of other governments, of firms, and of individuals.[5] In addition to command and control options, which might better be called direction of action, this range includes specifications of performance and provision of incentives. Within each of these broad categories there are, in turn, many specific alternatives. For any particular environmental quality problem it is likely that there will be available several possible implementation incentive systems from across the spectrum of possibilities.

This spectrum is illustrated in table 1. Under "Directives and Incentives" are listed examples from each of the broad categories,

"Direction of Action," "Specification of Performance," and "Provision of Incentives." The other major parts of an implementation incentive system, "Inspection, Monitoring, and Measurement," and "Sanctions" are also included in the table and the reader will observe that making a choice from the first column implies something about the choices to be made from the second and third columns. For example, the choice of a basic instrument involving specification of the treatment equipment that must be installed implies that monitoring means inspection for the installation, a yes/no situation. Sanctions, in turn, are geared to the "no" branch of that pair of possibilities and involve some penalty for failure. If, on the other hand, the government in question chose to specify treatment performance in terms of percentage removal of pollution prior to discharge, monitoring would involve measuring influent and effluent loads. Failure to comply, in this case, could be defined in either yes/no terms, or the degree of noncompliance could be taken into account. If the former is done, the problem becomes one of deciding what sanctions are to be applied for "no" cases. If the latter definition is chosen, it is sensible to think of sanctions defined per unit of noncompliance--e.g., cents per each pound of discharge above permit conditions--and these begin to look rather like effluent charges.

When incentives are provided without specification of results or of actions to be taken, the goal of monitoring is different. There is no standard to be met by the activity, but it is essential that the activity be charged (or subsidized) on an accurate basis, and monitoring must be concerned with the discouragement of misreporting (of discharges, for example). Because misreporting of discharges is analogous to

Table 1. Elements of Implementation Incentive Systems with Examples

Directives and Incentives	Inspection, Measurement and Monitoring	Sanctions
DIRECTION OF ACTION		
Specify characteristic(s) of raw material input No more than S% sulfur fuel, target-specific degradable material	Monitor input quality compliance/noncompliance extent of noncompliance	Penalties--fines or jail terms--for failure to comply
Specify production process Amount of thermal insulation in buildings, contour plowing in agricultural crop production, dry peeling in fruit and vegetable canning	Inspect process for compliance compliance/noncompliance	
Specify residuals modification and/or handling process Activated sludge sewage treatment plant, debris basins on construction sites, require householders to separate used newpapers and paper packaging from other solid residuals	Inspect equipment for compliance compliance/noncompliance	or
Specify product output characteristics Amount of lead in gasoline, amount of phosphate in detergents, standard number/sizes/designs of glass containers	Monitor product quality compliance/noncompliance extent of noncompliance	Penalties per unit of noncompliance where applicable
Specify locations or limitations on locations of activities Designate industrial districts, limit density of construction where infiltration capacity for septic tanks is less than a specified rate, prohibit surface mining on slopes above some steepness	Inspect locations compliance/noncompliance	

	Monitor activities compliance/noncompliance	Penalties—fines or jail terms—for failure to comply

Specify extent, timing, type of activity
Prohibit trucks on particular routes
during particular time periods, prohibit
all-terrain vehicles in environmentally
fragile areas, prohibit aerial spraying
of pesticides when wind is above some
speed, stagger work hours

SPECIFICATION OF PERFORMANCE

Specify allowed total quantity of residual
discharges per unit production
$\leq$ X kilograms of suspended solids per
ton of steel, $\leq$ Y grams of hydrocarbons
per vehicle-kilometer traveled

Monitor production and
discharge
compliance/noncompliance
extent of noncompliance

Specify allowed total quantity of residual
discharged per unit of time
$\leq$ Z kilograms of suspended solids per
day, $\leq$ T kilograms of particulates
per day

Monitor discharges
compliance/noncompliance
extent of noncompliance

Penalties—fines
or jail terms—for
failure to comply

or

Specify allowed concentration of residuals in
discharge
$\leq$ S mg/l of suspended solids in
wastewater

Monitor discharges
compliance/noncompliance
extent of noncompliance

Penalties per unit
of noncompliance
where applicable

Specify treatment performance
$\leq$ R% removal efficiency from wastewater
stream or stack gas

Monitor treatment process
influent and effluent
compliance/noncompliance
extent of noncompliance

(continued)

Table 1 continued

Directives and Incentives	Inspection, Measurement and Monitoring	Sanctions
Specify efficiency of conversion Auto mileage, appliance efficiency	Measure efficiency of conversion in operation compliance/noncompliance extent of noncompliance	Penalties--fines or jail terms--for failure to comply or
Specify ambient quality level to be achieved Annual or daily average concentrations of pollutants, concentrations of con-version products (ozone, dissolved oxygen)	Monitor AEQ (and related discharge[s] compliance/noncompliance extent of noncompliance	Penalties per unit of noncompliance where applicable
Create marketable rights to discharge T tons per day SO_2	Monitor discharges compliance/noncompliance extent of noncompliance	Penalties for failure to comply with right, or charge per unit of excess discharge
Create marketable rights to degrade ambient quality M mg/m^3 annual average concentration at point P	Monitor AEQ and related discharges compliance/noncompliance extent of noncompliance	

PROVISION OF INCENTIVE

Directives and Incentives	Inspection, Measurement and Monitoring	Sanctions
Charges per unit of residuals discharged X cents per kg of BOD_5, Y cents per 10^6 kilocalories, Z cents per kg of SO_2	Measure (audit) residuals discharge	Penalty for misreporting of charged or subsidized activity
Charge per unit of related input S cents per kg of sulfur in fuel, D dollars per kg of pesticide employed	Measure (audit) input use	
Charge per unit of related output P cents per kg of package weight, H dollars per year per auto horsepower	Measure (audit) output	

Charge on related activities	Monitor (audit) application of charge	
Surcharge on downtown parking fees		
Subsidy per unit of discharge reduction	Measure base and actual discharges	
Subsidy per unit of reduction on related input use	Measure base and actual use	Penalty for misreporting of charged or subsidized activity
Subsidy per unit of reduction on related output	Measure base and actual output	
Subsidy for related activities	Monitor performance of activity	
Treatment plant construction, agricultural practice upgrading		

Source: Modified from Bower, B. T., C. N. Ehler, and A. V. Kneese. 1977. "Incentives for Managing the Environment," Environmental Science and Technology vol. 11, no. 3, pages 250-254.

misreporting of income for tax purposes, it is possible to imagine that
the applicable sanctions could include both a penalty for the act of
misreporting itself (fraud) and a per unit penalty based on the size of
the "error."[6]

One form of implementation incentive system deserves brief special
mention. Marketable permits to discharge pollution or to add to the
ambient pollution load at certain points can usefully be seen as a
hybrid of the performance specification and economic incentive categories.[7]
The permit(s) held by a plant or other activity constitute a performance
specification. On the other hand, to the extent that they are in
fact legally marketable, and to the extent that a market for them exists
so that the holder can observe a non-zero price and thus calculate the
value of reducing (or the cost of increasing) discharges, the permits
imply an economic incentive analogous to the effluent charge. Monitoring
of such a system involves ascertaining compliance with the terms of
the permit(s); but again, sanctions can be imposed either for failure
itself or per unit of failure.[8]

Some other observations may be made on the basis of table 1.
First, there is no rule saying that in trying to achieve a particular
end a government must choose only one from among the many available
alternatives. In fact, actual systems, including the ones described
below in this book, usually involve combinations of directions, performance
specifications, and incentives (whether "up front" as an effluent charge
or buried in the sanctions for failure to take an action or meet a
performance standard). Second, there does not seem to be any necessary
general rule connecting the level of government making the choice (local,

state, regional, or national) and the range of choices open. In addition, it is possible for several levels to try simultaneously to affect the same problem through different implementation incentive systems. (This kind of overlap is, however, unlikely to be popular with those subject to the regulations, which is one reason why nationally uniform laws preempting a field for federal action are so frequently found in the United States.)

A related observation, however, is that a particular government's choice of implementation incentive system will be influenced by which aspects of the activities to be controlled are subject to that government's jurisdiction. This is illustrated schematically in figure 1. Particular incentive systems are logically applied to specific points within the activities and it is true that some specific actions will normally fall within the province of one jurisdictional level rather than another (e.g., zoning is normally a local government instrument; output specification for nationally-traded products will usually—though not always—be a system reserved to the federal government).

A fourth point to be made about implementation incentive systems is that some are more useful in the context of securing initial compliance (e.g., a directive to install certain equipment) and other systems are more obviously attuned to the problem of continuing compliance (e.g., specification of an allowed amount of discharge per day or per ton of output). In confirmation of this point, implementation incentive systems will often include the permitting process, which may be comprised of two parts. The first involves approval of the design of facilities which are state, regional, or national) and the range of choices open. In addition, it is possible for several levels to try simultaneously to affect the same problem through different implementation incentive systems. (This kind of overlap is, however, unlikely to be popular with those subject to the regulations, which is one reason why nationally uniform laws preempting a field for federal action are so frequently found in the United States.)

A related observation, however, is that a particular government's choice of implementation incentive system will be influenced by which aspects of the activities to be controlled are subject to that government's jurisdiction. This is illustrated schematically in figure 1. Particular incentive systems are logically applied to specific points within the activities and it is true that some specific actions will normally fall within the province of one jurisdictional level rather than another (e.g., zoning is normally a local government instrument; output specification for nationally-traded products will usually--though not always--be a system reserved to the federal government).

A fourth point to be made about implementation incentive systems is that some are more useful in the context of securing initial compliance (e.g., a directive to install certain equipment) and other systems are more obviously attuned to the problem of continuing compliance (e.g., specification of an allowed amount of discharge per day or per ton of output). In confirmation of this point, implementation incentive systems will often include the permitting process, which may be comprised of two parts. The first involves approval of the design of facilities which are

Figure 1. Loci of Possible Implementation Incentives for Water Quality Management

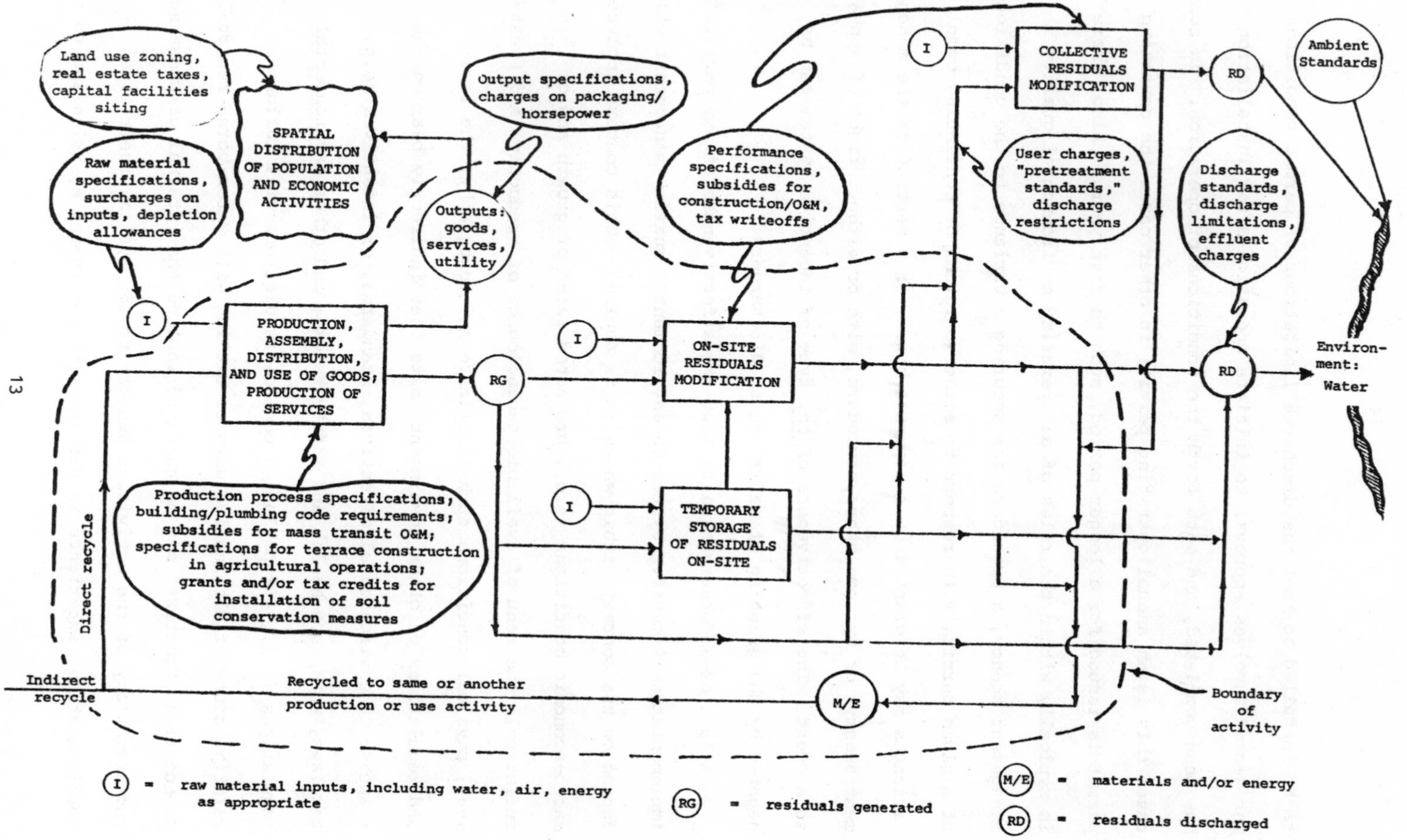

Ambient Standards

Discharge standards, discharge limitations, effluent charges

COLLECTIVE RESIDUALS MODIFICATION

User charges, "pretreatment standards," discharge restrictions

Performance specifications, subsidies for construction/O&M, tax writeoffs

ON-SITE RESIDUALS MODIFICATION

TEMPORARY STORAGE OF RESIDUALS ON-SITE

Output specifications, charges on packaging/horsepower

Outputs: goods, services, utility

SPATIAL DISTRIBUTION OF POPULATION AND ECONOMIC ACTIVITIES

Land use zoning, real estate taxes, capital facilities siting

Raw material specifications, surcharges on inputs, depletion allowances

PRODUCTION, ASSEMBLY, DISTRIBUTION, AND USE OF GOODS, PRODUCTION OF SERVICES

Production process specifications; building/plumbing code requirements; subsidies for mass transit O&M; specifications for terrace construction in agricultural operations; grants and/or tax credits for installation of soil conservation measures

Environ-ment: Water

Boundary of activity

Recycled to same or another production or use activity

Direct recycle

Indirect recycle

I = raw material inputs, including water, air, energy as appropriate

RG = residuals generated

M/E = materials and/or energy

RD = residuals discharged

13

to be installed to meet the discharge limitations or permit conditions.
The second involves approval to initiate operation after installation
has been completed, and sets forth the conditions of operation. In some
cases this is an annual operating permit; in other cases the operating
permit is issued for a longer period, such as five years. Also, there
is typically within the rules of an operating environmental quality
management agency, a procedure for securing a variance to the conditions
of a given permit, with respect to either original or continuing compliance.
Variances may be granted by a separate board, or a section of the management agency, or by some other procedure, with or without right of appeal
to a court. The effectiveness of _any_ type of incentive system can be
negated by the granting of variances to dischargers.

A last broad observation is that the effectiveness of any particular
implementation incentive system in environmental quality management will
depend on the context within which it is applied. This context includes:
macro-economic conditions (e.g., projected rates of growth and of
inflation); the menu of available technologies of production and treatment; physical conditions such as climate and topography; the laws
and regulations of other government units that indirectly bear on the
response to environmental directives, standards, or incentives (e.g.,
tax laws, workplace health and safety rules); and other environmental
regulations (e.g., air pollution control regulations will influence
relative costs of alternative ways of dealing with water-borne residuals).
In addition, "context" may be taken to include the internal circumstances
and psychology of the activities subject to the system: their size,
sophistication, experience in dealing with government agencies and so on.

In choosing an implementation incentive system for environmental policy, a government can and should take account of the context created by its directly or indirectly related policies. It should also be aware that its choice can, in the longer run, affect such contextual variables as available technology and sophistication of regulated activities. Thus, it is usually a mistake to take "context" entirely as a given, for this implies a short-run view of the problem.

EFFLUENT CHARGES

The last section provided a backdrop against which to set the particular type of implementation incentive system, effluent charges, of central interest in this book. In economic theory there exists an idealized effluent charge, the appeal of which seems to spill over into the practically attainable varieties of charges. Ideally, the charge per unit of residual discharge from a particular activity is equal to the sum of all the damages attributable to the last unit of that activity's discharge--when every activity has adjusted so that its marginal cost of residuals discharge reduction is equal to the effluent charge it faces. This situation represents a social efficiency optimum, the sum of costs of discharge reduction and damages from the remaining discharges being a minimum. It assumes, however, the availability and application of a great deal more skill and information than society has at its command. Most fundamentally, we cannot now measure in a persuasive way most of the damages attributable to pollution of any kind from any source.[9] If we could measure damages we would still be faced with an

immense computation job if we wished to find the socially optimal charge
set in any region. This would require a very large and complex model
including all relevant cost-of-discharge-reduction functions, characteri-
zations of the regional environment in its role as transporter, diluter,
and transformer of residuals discharges, and the set of damage functions
covering the region and categories of damages.[10] Such a model would
require that vast amounts of accurate data be available to the government
agency seeking the optimal charge. (Notice that because this optimal
charge set will in general entail different unit charges for each
discharger--not to mention each type of residual discharged--the possibility
of trial and error hunting for it through real-world experimentation
cannot be taken seriously.)

Faced with these obstacles to an ideal charge system, economists
and others have suggested a more practical variant--one in which charges
are used to meet exogenously chosen ambient environmental quality
standards.[11] In this limited context there also exists a socially
efficient charge, one that results in meeting the standard at least cost.
Again the optimal charge set will in general involve different charges
for different sources of any particular residual. Again, therefore,
finding the efficient charge set will require computation in a centrally
maintained model, including representations of the cost-of-discharge-
reduction functions for every source. Real-world trial and error would
only work in special circumstances in which discharge location (and
other characteristics) did not matter in determining the effect of one
unit of discharge on ambient quality.

Further relaxation of the conditions the effluent charge is to meet
produces systems that require less data and modeling virtuosity within the

agency imposing the charge. For example, it is at least possible to imagine a charge that is (for a particular residual) uniform over an entire region, being set by trial and error to achieve a given ambient quality standard. Such a charge would not meet the standard at least cost, however, and it would be an empirical question how the cost implied compared with that resulting from some other implementation incentive system, such as treatment plant specification or a discharge reduction (performance) standard.

Regardless of the obstacles to the application of optimal effluent charges raised by data availability and model complexity, it is possible to make one claim for any of the available charging systems: an effluent charge provides some incentive to every discharger to seek out or to try to invent and develop technology that uses the environment less intensively (e.g., production processes that generate less waste, treatment processes that achieve the same removal efficiencies more cheaply). It is impossible for practical purposes to predict the strength of this effect for a given charge or to calculate a dynamically optimal charge from this perspective, but it is nonetheless comforting to know that at least the incentive supplied operates in the right direction over time.

The record of adoption of effluent charges by governments has not, however, been an especially heartening one, and it is natural to ask why the potential advantages of having a charge, in one or another form, are so consistently seen as not worth it. Two reasons seem to be the ones most frequently offered. First, charges are said to be politically infeasible, a claim that in turn rests on three quite different concerns:

- Dischargers see a charge as merely adding to their costs, because not only will they have to pay for some level of discharge reduction, as under a directive or performance specification system, they will have to pay for their remaining discharges as well.

- Many environmentalists and politicians do not really believe that dischargers will act rationally. Rather, it is feared, the dischargers will simply pay the charge for the privilege of polluting excessively. (A more sophisticated version of this view is that the sources to which the charge applies will be too uninformed or too absorbed in other problems to find the optimal balance of removal effort and remaining discharge, or will be protected by monopoly market positions and able to pass on the charge more easily than react to it.) [12]

- There is also a feeling prevalent among environmentalists and their major allies on legislative staffs and committees that it is inappropriate to "put a price on" pollution and the resulting effects on natural environments and human health. [13]

The second major reason for nonadoption of charges is the fear that the system, even the simple uniform charge system, will be a legal and administrative nightmare. (The implication of this concentration on the administrative feasibility of charges is that systems of direct regulation are much simpler in this respect. But the history of the

United States' efforts over the past decade makes this implication difficult to accept.) The specific kinds of worries behind this general view include:

- finding legally defensible bases for setting charges;
- designing and operating a suffuciently tight monitoring system (i.e., do all discharges have to be monitored continuously?)

The special importance of the European experience, then, arises from the very fact of the existence there of effluent charges. How have the problems of political and administrative feasibility been solved? Do the European charge systems work? (A question that raises another: what is meant by "work"?)

IMPLEMENTATION INCENTIVE SYSTEMS
FOR WATER QUALITY MANAGEMENT
IN FRANCE AND GERMANY

It is the goal of the two major descriptive sections of this book to provide a complete picture of the complex systems for water quality management developed at the national level in France and at the "land" (or provincial/state) level in Germany.[14] On the basis of this picture the reader will be free to make his or her own assessment of the central issues raised above, and to decide what lessons the French and German experience hold for the United States. It will, however, be worthwhile to summarize here some important features of these systems in order to put the effluent charges in perspective. In addition, it will not be amiss to hazard some observations about political and administrative feasibility, to describe the difficulties of deciding how well the

systems work, and to suggest one assessment of the lessons for the

United States.

Both the French and German implementation incentive systems are

complex, with many important elements in addition to effluent charges.

Indeed, as has already been said, to describe these systems as "effluent

charge systems" is itself misleading. In both systems discharge standards,

embodied in one or another form of government-issued permit, are central

to the effort to maintain or improve water quality. Thus, in the

terminology of the first part of this chapter, the systems are at heart

based on specifications of performance.

This approach is reinforced both by extensive subsidization and by

effluent charges. Subsidies are available in both systems for the

construction of treatment plants (and for some inplant process changes);

and in the French system to reward subsequent performance of the plant

in residuals removal. To a significant extent these subsidies are

financed by the effluent charges. But since the program of subsidies

must be mapped out with fair certainty for several years at a time,

the level of charges must be set to provide sufficient revenue to fund

the anticipated subsidies (plus, where applicable, the cost of centralized

treatment activities, instream modifications, low flow augmentation,

and the administrative expenses of the responsible bodies). Thus, the

effluent charges faced by individual activities are not set at levels

meant to approximate marginal damages, to stimulate certain levels of

treatment, or to produce desired levels of ambient water quality. Rather,

they are designed to raise a given amount of revenue. To simplify just

a bit, the vector of the discharges of each activity is translated
into a scalar number, termed "population equivalents," by one or another
formula.[15] The revenue needs of the charging authority for the year
or period in question are then divided by the total number of chargeable
population equivalents to get a unit charge.

The difference between the French and German effluent charges and
the familiar models of the environmental literature is further accentuated
by the method used to come up with discharges (either in terms of BOD,
COD, etc., or as populations equivalents directly) for individual
activities. For most sources, chargeable discharges are estimated in
advance on the basis of rules of thumb applicable to all sources of a
particular type (e.g., tanneries, breweries, paper mills of specific types,
municipal treatment plants serving cities of particular sizes). Sources
that believe their waste streams to be cleaner than allowed for in the
rule of thumb can demand individual monitoring (paid for by the charging
authority or by the source, depending on outcome of the measurement);
but those activities doing worse than implied by the rule of thumb
pay only the rule of thumb charge and are only in trouble if they violate
(and are caught violating) the terms of their permits.[16]

This approach to charge setting and effluent monitoring is, of course,
consistent with the aim of the charges, which is to raise, in a reasonably
predictable fashion, revenue to be dispensed as subsidies or used for
regional projects. Indeed, in the circumstances, were there a significant
incentive effect from the charge in the short run, other things remaining
equal, the unit charge would have to be raised to allow meeting the
revenue subsidy commitment. This would be a perverse incentive, though not,

probably, a very important one, since from the single source's point

of view the increase due to his discharge reduction would reflect the

spreading of the revenue shortfall over all sources.[17]

What of political and administrative feasibility? As for the first,

it seems worth emphasizing both the importance of subsidies and the

method of instituting the systems over time. Because the charges paid

go largely into subsidies returned to dischargers, the sources in the

region end up in aggregate about where they would be without either

charge or subsidies but with only their permit requirements to meet.[18]

Individual sources and categories of sources can do better or worse

than this aggregate neutral result, and the description of the French

system below makes clear how peculiarities of the payment mechanisms

can accentuate these differences. (For example, municipal governments

receive subsidies for operating treatment plants but are not responsible

for collecting the effluent charges from their citizens. Thus, in

financial and, more important, political accounting, French municipalities

come off quite well. But in terms of full resource costs, by whomever

paid, it appears that users of municipal treatment plants come off less

well than similarly sized industrial sources. (See chapter 7 and chap-

ter 8.) Another key to acceptability appears to have been the very

gradual raising of charges to operational levels. The initial charges were

essentially trivial and thus sources had a chance to get used to the

mechanism before the payment even could be felt. A third factor,

relevant in the German context, is that the regional agency constructed

and operated facilities, particularly treatment plants, into which a

number of activities discharged directly.

On the administrative side, there are also two features of the
systems that seem especially important in making for feasibility. One
is the rule of thumb approach to estimating discharges subject to the
charge. This relieves the authorities involved of the necessity for
auditing the reported discharges of every source. (Of course, it is
necessary for the relevant authority to have some credible monitoring
system for permit compliance and the results of this could be transferred, if desired, to the charge area.) Second, the level of the unit
charge itself is set, not through some complex modeling process or through
a trial and error system (with its attendant problem of seeming arbitrari-
ness), but in a well-defined and (reasonably) clear calculation based
on revenue needs. This process leaves much less room for challenge by
affected sources. Beyond these observations there is the much "softer"
but undoubtedly real matter of the different traditions of government
direction and control, and the different attitudes toward, and
possibilities for, legal challenges to government actions in France and
Germany as opposed to the United States. Pursuing this idea would very
quickly take this introduction beyond its depth and competence. It
does, however, constitute a general warning about the applicability of
European experience to U. S. problems, a matter that is pursued with somewhat more specificity below.

Do the French and German mixed implementation incentive systems
work? More specifically, do their effluent charge components work?
The short and easy answer is yes. That is, the systems do actually
operate, and impressionistic evidence strongly suggests that the quality
of the managed water courses has both improved since they began operating

and is generally higher than it would be without them. It is, however, difficult to put numbers on these impressions.[19]

At the level of the individual discharger there has either been no systematic sampling over time (true in France) or, if such sampling has been done, the records are not available to outside researchers (in the Genossenschaften). An alternative approach is to compare the levels of effluent charges with the costs of building and operating waste treatment plants (expressed per unit of pollution removed from the influent stream). This is done for the French charges in part II of this book. The comparison indicates that the charge levels so far employed do not offer, by themselves, an incentive to reduce discharges-- at least not by conventional treatment routes.[20]

A similar lack of data bedevils any attempt to quantify the improvement in ambient water quality that might be attributed to the implementation systems. Exactly these problems also plague attempts to assess the effectiveness of U. S. environmental laws and regulations: lack of measurements covering space, time, and quality indicators; inaccurate measurement techniques; difficulty of comparing long strings of varying measurements over time.[21] Further, whatever evidence of improvement does exist must be viewed in relation to each system as a whole and cannot be attributed to the use of effluent charges.

On the other hand, one might say that the French and, to some extent, the German state systems are quite similar to the U. S. system as it has been cobbled together over the past couple of decades--except for the effluent charges. Can it be said then that the European experience has been better than this country's, the difference being attributable to the use of charges? It does not seem possible to say this without a major effort at data gathering, analysis and interpretation. Such information as is readily available suggests that there is no striking difference in effectiveness; that perhaps more ambitious U. S. goals and their clear linking to ambient water quality, and the more litigious system generally, encourage the impression of at least partial failure and set critics casting about for better methods.

All this further suggests that the lessons of the European experience for the United States are rather limited. It is clear that there is nothing necessarily infeasible about combining standards, charges, and subsidies, and making charges work--at least to the extent of collecting money. That is, the political and administrative feasibility of instituting a charge per unit of discharge has been demonstrated. On the other hand, the French and German systems tell us very little of relevance to the model effluent charge systems usually proposed. The very features that seem to make for feasibility differentiate these systems from those in which a charge is used to stimulate directly the attainment of a desired ambient quality standard.[22] These features, to summarize, are:

- the revenue (subsidy) need basis for the charges;
- the relatively low levels of the charges (though this is changing fairly rapidly);
- the rule of thumb method of charge assessment.

In the longer run a different conclusion may prove warranted. To the extent that: 1) incentive effects are heightened by gradual substitution of measurement for rule of thumb assessments; 2) charge levels rise because of the need to subsidize more and larger treatment and

sewerage projects; and 3) monitoring is put on a sustained and systematic
basis; it may begin to be possible to see the dynamic effects referred
to early in this chapter--the encouragement of technological developments
in the direction of less residuals-intensive production processes,
cheaper or more efficient treatment methods, and so forth. Over this
longer horizon one might see significant differences open up between
the United States and those European countries with effluent charges as
part of their implementation incentive systems for water quality management.

A BRIEF PLAN OF
THE BOOK

Part II of the book deals with the French national implementation
incentive system. First, in section 1 of part II, chapters two through
six, the water management system in France is described, with primary
emphasis on water quality management and with a careful description of
the relationship between the economic and regulatory aspects of the system.
Second, in section 2 of part II, chapters seven through ten, the effluent
charges system itself is described, giving particular attention to the
process of determination of the levels of the unit charges on various
substances. Chapter eleven contains some conclusions concerning the
role and effects of effluent charges in France.

In part III the book takes up the implementation incentive system
used by the river basin authorities (Genossenschaften) of the Nordrhein-
Westfalen state of West Germany. Chapter twelve provides some background
on the Ruhr area of the Federal Republic of Germany. Chapter thirteen
describes the governmental structure for water quality management in the

area of the Ruhrverband/Ruhrtalsperrenverein, and chapter fourteen
describes in detail the procedures for computation of the unit charges
on discharges by those associations. Chapter fifteen provides details
on computation of unit charges by the Lippeverband, another Genossenschaft
in the Ruhr area. The final chapter, sixteen, presents some conclusions
on the role of effluent charges in the Genossenschaften areas.

NOTES

[1]Schultze, Charles, _Public Use of Private Interest_ (Washington, D. C.,
Brookings Institution, 1977).

[2]For discussion of this point in relation to effluent charges and
marketable permits, see respectively: Russell, C. S., "What Can We
Get from Effluent Charges?" _Policy Analysis_ vol. 5, no. 2 (Spring 1979);
Russell, C. S., "Environmental Policy and Controlled Trading of Pollution
Permits," _Environmental Science and Technology_ (in press).

[3]An introductory catalog of studies giving substantial attention to
these systems would include:

> Kneese, Allen and Blair T. Bower. 1968. _Water Quality Manage-
> ment: Economics, Technology, Institutions_ (Baltimore:
> Johns Hopkins University Press for Resources for the
> Future).
> Irwin, Will. 1971. _Wastewater Treatment Assessments, Legal
> and Administrative Aspects_ (Washington, D. C.: Environ-
> mental Law Institute).
> Johnson, Ralph and Gardner Brown. 1976. _Cleaning Up Europe's
> Waters_ (New York: Praeger).
> Organization for Economic Cooperation and Development. 1980.
> _Pollution Charges in Practice_ (Paris).

[4]The West German national law (Abwasserabgabengesetz) of 1976 does
involve the imposition of effluent charges that conform more closely
to one of the models used implicitly or explicitly by charge advocates.
But there is as yet no experience with this law because the imposition
of charges does not begin until 1981.

[5]Implementation incentives and implementation incentive systems are directed toward individual activities, but the response of an activity is conditioned by whether or not the activity is a single plant firm or one plant of a multi-plant firm. In this connection it is useful to point out how misleading can be the use of the term, "firm." Traditional economic theory uses the word in the sense of the individual entrepreneur, the owner of a single plant. This has long since ceased to mirror reality, with the multi-plant firm or company having become common. If the individual activity is one plant of a multi-plant firm, the decision(s) relating to adoption of measures to reduce discharges from that plant are taken within the context of the entire company. Factors which are likely to differ between single plant and multi-plant firms are: (a) availability of capital; (b) availability of technical advice; and (c) availability of legal advice. In a multi-plant firm--if it is a major residuals generator--there is likely to be a special department of environmental control, which can provide specific information on technological options and their costs. There is also likely to be an internal legal department with experts on environmental law on the staff.

[6]It is worth noting that this analogy between self-reporting of income and of pollution discharges cannot be pushed too hard. The economy runs on paper, and transaction records are ubiquitous, implying that individuals and firms will be hard put to conceal incomes. Such concealment is really only a consistently workable option in the purely cash economy such as that in which the street peddler of legal or illegal goods operates. In much of the economy one person's (or firm's) income is another person's (or firm's) deductible expense, so record keeping requirements carry their own implementation incentive system. The situation is quite different for pollution discharges, first, because there is no transaction and no other party (except the government agency) with an interest in record keeping. Second, even if effluent charge payments are tax deductible, it is still likely to be cheaper (sanctions aside) to avoid paying them in the first place.

[7]For a good clear discussion of marketable permits, see Thomas Tietenberg, "Transferable Discharge Permits and the Control of Stationary Source Air Pollution: A Survey and Synthesis," (Waterville, Maine, Colby College, 1979, unpublished).

[8]There is a special problem where the permits cover the right to reduce ambient environmental quality at particular points in the region. If more than one source of pollution is involved, monitoring must involve discharges rather than the actual permit conditions, and the translation of discharges into ambient quality decrements by mathematical modeling is at best an uncertain art.

[9] Work is and has been underway to improve on this situation but it is unlikely that the statement in the text will be outdated any time soon. See A. M. Freeman III, The Benefits of Environmental Improvement (Baltimore, Johns Hopkins University Press for Resources for the Future, 1979).

[10] It seems to be an open question whether, when such pervasive effects as those arising from visibility, odor, etc., are included, the job is even in principle manageable.

[11] The discussion that follows is based on C. S. Russell, "What Can We Get from Effluent Charges?" Policy Analysis vol. 5, no. 2 (Spring 1979), pages 155-180.

[12] Fundamentally this concern assumes that activities in a market economy do not respond to changes in factor prices, i.e., are not cost minimizers. The responses of industrial and even residential activities to the sharply increased costs of energy in the latter half of the 1970s probes the validity of this view.

[13] For striking evidence on this point from interviews with environmentalists, Congressional staff members, and business lobbyists, see Steve Kellman, "Economic Incentives and Environmental Policy: Politics, Ideology, and Philosophy" (Cambridge, Mass., Kennedy School, Harvard University, 1980, unpublished paper).

[14] There are four primary sources of information which are the basis for the descriptions and analyses in Parts II and III of this volume. One consists of actual records and budgets of the French river basin commissions, particularly the Seine-Normandie, and the Genossenschaften in the Ruhr. The second is the personal experience of the native authors as professionals associated with water resources management in both countries. The third consists of legislative documents of various types. The fourth consists of discussions with various of the actors in both the public and private sectors.

As noted above there now exists in Germany a national water pollution control law of which one feature is a system of effluent charges. These charges will be levied beginning on 1 January 1981 and are coupled with discharge permits. There is apparently not a tie to a subsidy system, though the unit charge levels are linked to the application of certain approved technologies for discharge reduction. See Horst Seibert, "Practical Difficulties in Guiding Environmental Use through Prices," University of Mannheim Discussion Paper 141/80 (English translation available from Resources for the Future, Washington, D. C.).

[15]These expressions are of the general form:

$$PE = 1/\alpha \ BOD + 1/\beta \ COD + 1/\gamma \ TSS + 1/\delta \ Toxics + \ . \ . \ .$$

where	PE	=	population equivalents
	BOD	=	a measure of biochemical oxygen demand of the waste stream
	COD	=	the chemical oxygen demand of the waste stream
	TSS	=	the total suspended solids load of the waste stream
	Toxics	=	a measure of the toxicity to aquatic organisms of the waste stream
	$\alpha, \ \beta, \ \gamma, \ \delta$	=	constants reflecting the strength, in terms of each residual, of "normal" household wastewater.

[16]In France, most of the major discharges have now been measured.

[17]Since all sources see this effect in the same light, however, it is possible that ex post each might find that action to reduce discharges resulted in no net change in total charge paid.

[18]A somewhat different fee-subsidy system is being suggested for the control of air pollution in Philadelphia. James D. Smith, "Emission Fee/Subsidy Concept," unpublished note, City of Philadelphia, Air Management Services, July 1980.

[19]Some evidence from a number of Organization for Economic Cooperation and Development (OECD) countries, including France but not Germany, is brought together by Jean-Philippe Barde, Gardner Brown and Pierre F. Teniere-Buchot, in "Water Pollution Control Policies Are Getting Results," Ambio vol. VII, no. 4 (1979) pages 152-159.

[20]See also the evidence on this point in Barde, Brown, and Teniere-Buchot, "Water Pollution Control Policies Are Getting Results."

[21]See, however, the interesting descriptive material on specific water bodies in U. S. Environmental Protection Agency, National Accomplishments in Pollution Control (draft report, Washington, D. C., U. S. EPA, 1980).

[22]The French system has been moving, and continues to move, in the direction of ambient-based charges, e.g., the Ambient Water Quality Objectives Policy, with charges and subsidies modified to reflect achievement or nonachievement of ambient water quality standards by subarea. See chapter 3 and chapter 11.

PART II

WATER MANAGEMENT IN FRANCE, WITH SPECIAL

EMPHASIS ON WATER QUALITY MANAGEMENT

AND EFFLUENT CHARGES

Rémi Barré and Blair T. Bower

33

PART II

Table of Contents

Section I

PRINCIPLES, ORGANIZATION, AND FUNCTIONING OF THE
WATER MANAGEMENT SYSTEM IN FRANCE: THE RELATIONSHIP
BETWEEN ITS ECONOMIC AND REGULATORY ASPECTS

The objective of this section is to present the overall organization of the water management system in France. This system has two
basic characteristics. The first characteristic is the use of both
regulatory-legal controls and economic-financial instruments.[1] This
means that there exist at the same time: (1) procedures for authorizing withdrawals and discharges; (2) discharge standards; (3) ambient
water quality standards; and (4) effluent charges and water intake
charges.

This duality is expressed in the "White Book of Water in France"
which states that:[2] "The regulatory mechanisms of the market economy
can bring a solution to the problems (of water management)...and
the effluent charge system in the Basin Agencies[3]...comes from this
approach." But the report also states: "The economic approach can
only bring partial solutions...economic analysis cannot define
objectives, nor can economic stimulus be a sufficient means of
action...This is why a regulatory system is necessary...which needs
to be clear and coherent with economic action...." The report
concludes: "This shows the necessity of a permanent dialogue
between the local political institutions--communes, departments,
regions--and the Basin Organisms."

This duality also appears in the Law of 1964, the fundamental law on water in France, in which articles 2 to 6 and 40 delineate the regulatory system and the setting of discharge standards (norms) and articles 12 to 14 delineate the establishment of the effluent charge system and the basin agencies system.

The second basic characteristic is the complex relationship among the territorial entities, each represented by specific institutions or actors. On the regulatory side of water management, the major entities are the national level and the departmental level.[4] On the economic-financial side, the "Basin"[5] is the main entity.

The objective then, is to present the overall principles, organization, and functioning of the water management system in relation to these two basic characteristics.

FOOTNOTES AND REFERENCES

[1]This is also the case, to a greater or less extent, in The Netherlands, Denmark, Japan, Hungary, and the Federal Republic of Germany.

[2]La Documentation Francaise, Paris, 1974, pp. 100-101.

[3]Often referred to more simply as, "Agences de Bassin," or even "Agences."

[4]There are 91 departments in metropolitan France. These are areal units of government.

[5]There are 6 "Basins" in France. Each contains one or more of the major rivers of the country. The boundaries of the basins correspond roughly to the boundaries of the watersheds of those rivers.

Chapter 2

PRINCIPLES, POLICIES, AND GENERAL ORGANIZATION
FOR WATER MANAGEMENT

THE BASIC TEXTS, LAWS, AND DECREES: THEIR
RELATIONSHIPS TO POLICIES AND ACTIONS

The Civil Code and the Rural Code

These codes, dating back to the early nineteenth century--the time
of Napoleon I--define the rights and duties of land owners regarding
the waters which are on or adjacent to their properties. The Rural Code
gives police powers on water to the administrative authority, and defines
the limits of the property rights and obligation of maintenance. These
codes apply to "domainial" (navigable or floatable) as well as to "non-
domainial" (non-navigable or non-floatable) waters.[1] Article 434-1
of the Rural Code states, "whoever has discharged products that can
modify the ambient water environment can be condemned." This article
is still often used for legal action, primarily by fishermens' associations.

The Code of Public Health and
Departmental Sanitary Regulations

These define protected areas around intakes for drinking water,
norms for wastewaters discharged into the sewers, and general conditions
for sewage systems. They have long been, and still are to a certain

extent, the basis for checking and controlling new developments which do not belong to the category of "classified establishments," which is described in the next section.

Departmental sanitary regulations consist of a set of prefectoral[2] decrees, based on the general framework of the Code of Public Health, which allows taking into account the specific problems of each department, including water and sewage problems.

The Law of 19 December 1917, Reformulated by the Law of 19 July 1976, Known as the "Classified Establishments Law"

This is the fundamental law in France for the control of nuisances, including noise, air pollution, and hazards, as well as water pollution. This law establishes a specific procedure for authorization and control and applies to more than 400 types of activities, which are termed "classified establishments." These activities are divided into two categories, depending on the size and nature of the activity. The first is comprised of those activities which must only make a declaration to the Prefect and to the Classified Establishments Service prior to starting operations or modifying operations, provided that the activity meets other regulations, e.g., sanitary regulations and standards resulting from the decrees of 23 February 1973 and 13 May 1975, and departmental ambient water quality objectives. There are about 500,000 establishments in this category (category D). The second is comprised of those activities which must apply for a permit prior to starting or modifying their activities. This involves: acceptance of stipulations of the measures to be taken to prevent or reduce disamenities; holding of a public

enquiry; and preparation of an environmental impact statement. There

are about 50,000 establishments in this category (category A).

The law established the "Superior Council on Classified Establish-

ments" and the "Classified Establishments Service." The former

consists of thirty (30) individuals from: the various national

ministries; the Classified Establishments Service; chambers of

commerce, industry, and agriculture; and environmental associations.

The Council has two roles: (1) to advise the Minister of the Environ-

ment and Quality of Life on all legislative actions; and (2) to give

advice on major projects which require an authorization, e.g., nuclear

power plants. For example, the Council met three times in 1978.

The Classified Establishments Service (SCE) is formally

linked to the Ministry of the Environment and Quality of Life (Ministere

de l'Environnement et du Cadre de Vie, hereinafter referred to simply

as the Ministry of the Environment), but operates closely with the

Ministry of Industry. It is organized into seventeen "regional"

services, defined by combinations of departments. The seventeen

regions do not correspond to the twenty-two regions into which the

country has been divided (see figure 5). The functions of the SCE

are: (1) to make and implement the recommendations and stipulations

for existing and new plants belonging to Classified Establishments

of category A, prior to authorization; (2) to review and keep up-to-

date the nomenclature of the classified establishments;[3] and (3) to

define, for each activity, "normal conditions" for operation and

discharge of residuals. The procedure for authorization and control

was first established by the decree of 21 September 1977. With
respect to the first function, the inspectors of the SCE have a permanent and absolute right to enter, inspect, and take samples in
classified establishments whenever they wish.

About 50,000 establishments operate with an authorization from
the SCE. In 1978, 2,312 authorizations were issued, either for
completely new plants or for plants which had modified their operations; 1,217 of the authorizations were for industrial plants.

The SCE is traditionally composed of civil servants belonging
to the Corps of Mines. The personnel of the SCE consists of 917
individuals, including 377 inspectors. On the average, each inspector
is responsible for 250 establishments operating with authorizations.
In an average year each inspector: files 25 authorization procedures,
9 of which are for totally new operations; is responsible for 14.4
official complaints; issues 4 injunctions compelling establishments to
take action; and issues 0.7 sanctions.

The Law of 16 December 1964, Known as the "Law on Water"

This law, "relative to water distribution and pollution control,"
sets the general framework for modern water management in France. Its
general objective is "pollution control and water regeneration in
order to meet and reconcile the requirements concerning: drinking
water supply and public health; agriculture, industry, transportation
and other human activities of general interest; biological life in
waters and especially fish life, as well as recreation, water sports,
and site protection; and conservation and flows of water." The law

applies to all types of discharges, having all kinds of effects, for
all kinds of waters--including underground waters and sea waters--
within the territorial limits.

Various articles of this law, and associated decrees, establish
the following policies, actions, institutions, procedures. One, the
National Ambient Water Quality Inventory is instituted (article 3 and
decrees of 2 September 1969, 16 October 1970, 11 March 1971, 26 September 1975, and 2 February 1976). The Inventory consists of measuring
ambient water quality by at least 13 water quality indicators: (a) at
1,228 points, with sampling occurring at each point a few times a
year every five years; and (b) at 100 points a few times a year every
year. Measurements at all 1,228 points were made in 1971 and 1976.

Two, authority is provided to regulate the use and sale of
certain products or materials which can result in water pollution
(article 6.2 and decrees of 15 July 1971 and 28 December 1977). For
example, detergents[4] and organophosphate pesticides have been banned
under this regulatory authority.

Three, the Ambient Water Quality Objectives Policy is promulgated
(articles 3, 4, and 6.4, and decrees of 18 December 1970, 29 July 1971,
and 19 March 1978). This policy consists of: (1) defining criteria
for the quality of water for certain uses, e.g., drinking water,
recreation, swimming; (2) defining a predominant type of use for
each section of a river within a department; and (3) deciding on
the effluent standards to be applied in each of those sections.

Four, wastewater discharge regulations are specified. Article 6.1
and the decrees of 25 September 1970, 23 February 1973, 13 May 1975,
and 14 January 1977 relate to the procedures for establishing local

standards and for authorizing discharges. Articles 6.3 and 9, and
the decrees of 15 December 1967, 22 January 1973, 13 March 1973, and
28 October 1975, relate to the procedures for sampling and analysis
of discharges.

Five, the basin committees of the basin agencies are created
(article 13 and decrees of 14 September 1966 and 28 October 1975);
the administrative boards of the basin agencies are created (article
14 and the decree of 14 September 1966); and the system of water intake
charges and effluent charges is established (articles 14, 14.1, 14.2,
and 18, and the decrees of 24 October 1967, 5 January 1970, and
28 October 1975).[5]

Other Laws and Decrees Having Impacts on Water Management

Several other laws and decrees have varying impacts on water
management.

- The law of 10 July 1976, Law on the Conservation of Nature,
  establishes the environmental impact statement procedure.
  This procedure plays a role in the discharge authorization
  process.

- The decree of 22 January 1968 states the role and the
  functioning of the Hygiene Council of each department
  which also has a role in the discharge authorization
  procedure.

- The decree of 19 November 1969 specifies the process prior
  to the issuing of a prefectoral decree.

• The decree of 1 August 1905 establishes the Public Enquiry
and Administrative Conference procedure.

Various decrees--from 1972 to 1977--establish the "sector contracts" between the national government and certain heavy-polluting
industrial sectors, for which sectors the necessary investments to
reduce discharges to water bodies would be especially high. Sectors
such as pulp for paper, beet sugar processing, and five subsectors
within the food processing industry, have committed themselves to a
program of reducing discharges, and hence receive special grants which
amount to about 10% of the capital investment in facilities to reduce
discharges.

Environmental legislation of the European Economic Community
was launched after the "Statement of Principles" of the Community
Council on 22 November 1973, based on article 2 of the Treaty of Rome
establishing the European Economic Community. This legislation, as it
relates to water, aims to control or ban certain toxic materials and
to define water quality criteria in relation to type of use of water.
These criteria will be used in the ambient water quality objectives
program.

Concluding Comment

Figure 2 shows how the above cited laws, codes, and decrees
interrelate to build five policies for water quality management.
Figure 3 shows how these five policies interrelate to build the
authorization and financing procedures which are the means for
inducing actions to improve ambient water quality.

Figure 2. From Eight Pieces of Legislation to Five Policies: Water Management in France

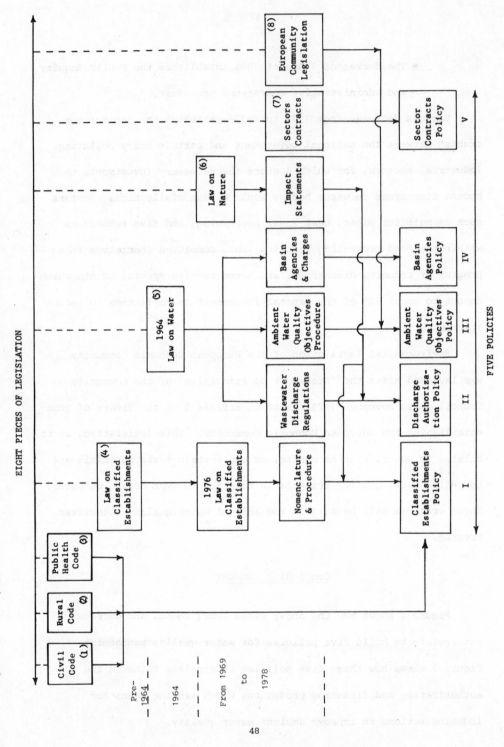

EIGHT PIECES OF LEGISLATION

FIVE POLICIES

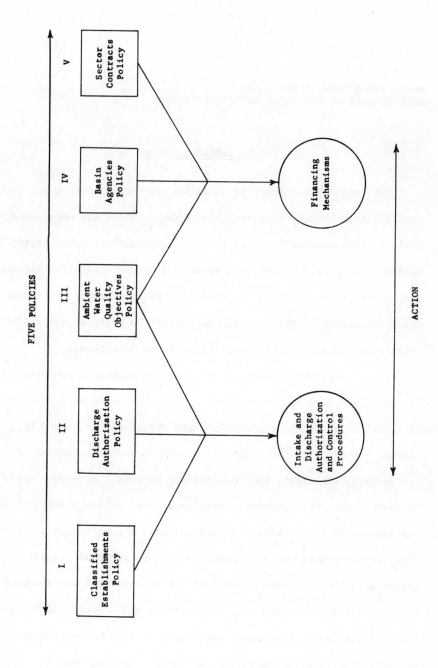

Figure 3. From Five Policies to Action: Water Management in France

FIVE POLICIES

| I | II | III | IV | V |

Classified Establishments Policy

Discharge Authorization Policy

Ambient Water Quality Objectives Policy

Basin Agencies Policy

Sector Contracts Policy

Intake and Discharge Authorization and Control Procedures

Financing Mechanisms

ACTION

BASIC MECHANISMS, ACTORS, AND
ORGANIZATION OF THE WATER MANAGEMENT SYSTEM

The Territorial Units

The national territory is administered through four areal units: regions, departments, communes, and basins. There are twenty-two regions, each comprised of two to seven departments. Each region is headed by a prefect of region,[6] appointed by the national government, and by a regional council,[7] which elects its president. This council has, practically, little power. Basically the regional government operates to coordinate the activities of the departments.

The 91 departments--each comprised of a number of communes-- average about 5000 km^2 in area, except for the Seine department (City of Paris) which is about 150 km^2. Each department is headed by a prefect appointed by the national government and representing it. The prefect, who issues the "prefectoral decrees," has--very often in fact--the real administrative, regulatory, and police powers, because the department is the basic territorial unit for virtually all national policies and programs. This power can be, to a certain extent, shared with the directors (heads) of the departmental divisions of the ministries of Environment, Agriculture, and Health[8] which represent their ministries in the department. Each department has a general council, also elected indirectly, as is the regional council. The general council has mainly a consultative role and decides about part of the public investments within the department. This is the level at which virtually all the authorization procedures with respect to

intakes and discharges operate. It is also the basic level for
implementation of the ambient water quality objectives policy.

There are 36,394 communes, ranging in population from zero to
several thousand. Each commune is headed by an elected municipal
council, which then elects the mayor. The council: initiates decisions
relating to certain public investments--including municipal sewer
systems, sewage treatment plants, and storm water runoff systems;
operates the sewer system and sewage treatment facilities; runs--or
has a private water distribution company run for the commune--the
water supply and water distribution system; has a consultative role
in the authorization of direct water intakes and wastewater discharges;
and has--to a certain extent--some police powers. However, a prefectoral authorization is usually needed for any significant municipal
decision.

There are six basins, whose boundaries do not coincide with
those of the regions or of the departments. The basins are the areal
units for actions of the basin agencies. These actions include
defining the levels of water intake charges and effluent charges,
establishing criteria for making investments to improve water
quality, and certain water-related planning activities. There is now
a tendency to subdivide, more or less informally, each basin into
subbasins in which closer contacts can be established with local and
departmental levels.

The regions, departments, and basins are shown in figures 4 and 5.

Figure 4.   The Six Basins and Twenty-Two Regions of France

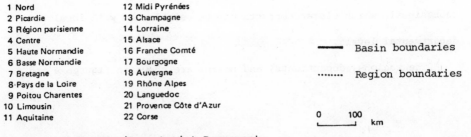

Regions

1 Nord	12 Midi Pyrénées
2 Picardie	13 Champagne
3 Région parisienne	14 Lorraine
4 Centre	15 Alsace
5 Haute Normandie	16 Franche Comté
6 Basse Normandie	17 Bourgogne
7 Bretagne	18 Auvergne
8 Pays de la Loire	19 Rhône Alpes
9 Poitou Charentes	20 Languedoc
10 Limousin	21 Provence Côte d'Azur
11 Aquitaine	22 Corse

———— Basin boundaries

········ Region boundaries

0   100
└─────┘ km

Source: S.P.E.P.E. (Secretariat Permanent
   pour l'Etude des Problemes de l'Eau)

Figure 5. The Ninety-one Departments and Twenty-two Regions of France

Source: adapted from:
    Principaux résultats du recencement de 1975, <u>les collections de</u>
    <u>l'INSEE</u>, n° 238, série D-52, Septembre 1977.

The Financial Flows

A critical component of water quality management in France involves the flows of money to finance sewage treatment plants and sewer systems; only industrial activities and municipalities can build and operate treatment plants. The resulting financial flows are shown in figure 6. For example, for municipal sewage treatment, capital costs of facilities are financed through: (1) loans and grants from the relevant basin agency; (2) the municipal budget; and (3) grants from regional and departmental budgets, which in turn come primarily from the national budget. Operation and maintenance (O&M) costs of facilities are obtained from: (1) the sewage tax; (2) the premium;[9] and (3) the superpremium,[9] in cases where good efficiency of operation is achieved in the sewage treatment plant.

Three points with respect to these financial flows relating to municipal sewage treatment merit emphasis. One, the sources of funds to cover capital and O&M costs appear to be clearly partitioned, but in fact they are not. For example, the sewage tax can well cover part of capital costs of facilities and the superpremium can go into the general municipal budget.

Two, households pay the effluent charge--via their water intake bills--based on assumed residuals generation, not discharge. However, the operator of the sewage treatment plant receives a premium based on the difference between the inflow to and the actual discharge from the treatment plant.[10]

Three, the effluent charge and the sewage tax are imposed on water intake. Hence, there is no incentive whatsoever to discharge

Figure 6.   Financial Flows Relating to Sewage Treatment

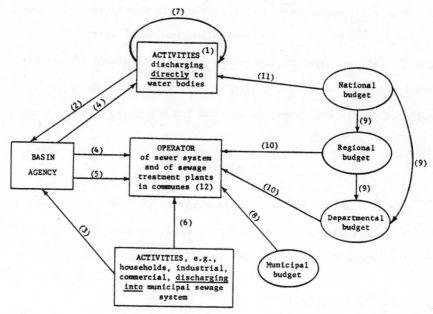

(1) That is, activities not discharging into a municipal sewage system.
(2) Effluent charge on actual discharge by individual activities other
     than municipalities. For municipalities payment is made based on
     generation with a refund--premium plus superpremium--based on pro-
     portion of generation not discharged.
(3) Effluent charge on generation, paid through the water distribution
     companies as addition to water intake charge. Generally the charge
     amounts to between 1/5 and 1/10 of the sewage tax.
(4) Loans and grants for part of the capital cost of sewage treatment
     plants only.
(5) Premium based on amount of generation not discharged. It is an
     operating cost subsidy. In the case of good efficiency in the
     operation of the plant, there is also a superpremium, based on plant
     load factor and efficiency.
(6) Local (municipal) sewage tax for capital costs of both sewers and
     sewage treatment plant(s). This is in reality a direct user charge.
(7) Capital and operating costs of the sewage treatment plant paid
     by the activity itself.
(8) Capital costs of the sewer system and of the sewage treatment plant
     paid directly out of the municipal budget. (Costs of storm water
     runoff system also come directly from the municipal budget.)
(9) National grants to regions and departments; regional grants to
     departments.
(10) Grants from regions and departments for some part of capital costs
     of sewer system and of sewage treatment plant.
(11) Grants and loans from national government for some part of capital
     costs, according to the "sector contracts" between the national
     government and certain industrial sectors.
(12) Usually the operator is a municipality or a set of municipalities.

less, for a given amount of water intake. In particular, the system
is disadvantageous for an activity using municipal water for its
garden, i.e., not discharging into the sewage treatment plant, and
advantageous for activities, such as a commercial laundry, which
would not pay an additional amount for its discharge of chemicals.

The "Actors" of the Water Management System

Figure 7 shows the overall organization for water management in
France. The following paragraphs provide details on those "actors"
for which their functions are not readily apparent.

Actors at the National Level

There are six bodies at the national level involved in water
management.

The Water Service.   The Service is part of the Ministry of
Environment and Quality of Life, in the Prevention of Pollution and
Nuisances Administration (DPPN). This small organization has a key role
as coordinator of policies at the national level. Its missions are:

- to prepare the work and be technical advisor to the
  Interministerial Mission on Water, the National Committee
  on Water, and the Scientific Committee on Water;
- to participate in the activities of the basin committees,
  the boards of administration of the basin agencies, the
  delegated basin missions, the National Committee on Water,
  and the Commission on Water for the Plan;[11]
- to prepare decrees concerning the 1964 Law on Water;

- to have the administrative supervision of the basin agencies and the delegated basin missions; and

- to organize the data bank on water and to manage the national water pollution inventory.

The National Committee on Water. This committee has sixty members chosen from among civil servants, elected officials, and water users. It has an advisory role and must be consulted about large-scale regional developments and problems common to several basin agencies.

The Interministerial Mission on Water. The Mission meets twice a year. A subgroup of the Mission, consisting of civil servants representing their respective ministries, meets twice a month. The Mission acts as a coordinator among all ministries involved, including the Ministry of Economics and Finances, and provides a link--in both directions--between technicians and the national government. It drafts laws and decrees concerning water. It plays an active role in defining national policies and in reviewing the multi-year programs of the basin agencies.

The Commission on Water for the Plan. The Commission is established every five years on the occasion of the delineation of the national economic and social development plan.

The House of Deputies and the Senate. These are the elected legislative--and controlling--bodies, with the Senate having only a consultative role. To these bodies are submitted the yearly budgets, including those of the basin agencies. However, the Senate has close contact with the general councils which vote the departmental budgets.

Figure 7. Overall Governmental Organization for Water Management in France

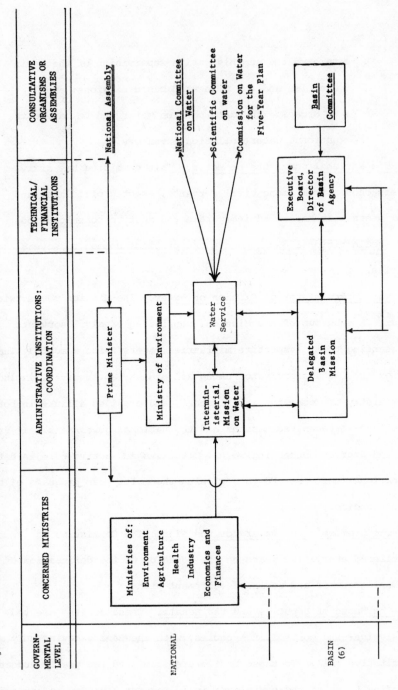

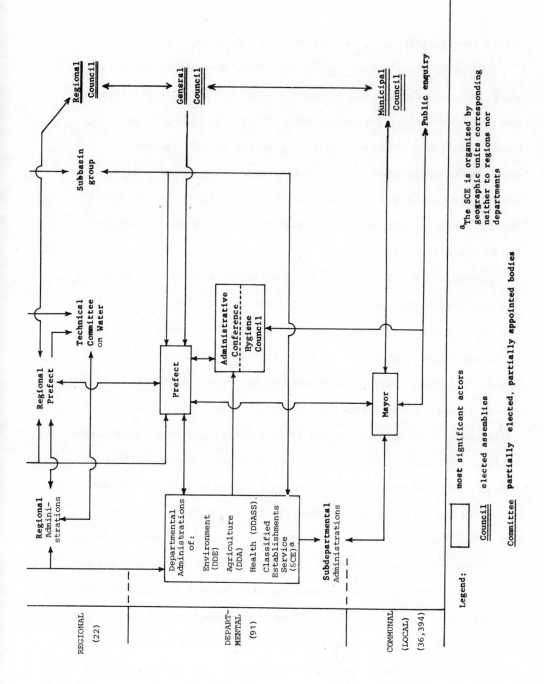

Legend:

most significant actors

Council   elected assemblies

Committee   partially elected, partially appointed bodies

[a]The SCE is organized by geographic units corresponding neither to regions nor departments

REGIONAL (22)

DEPART-MENTAL (91)

COMMUNAL (LOCAL) (36,394)

Regional Council

General Council

Municipal Council

Public enquiry

Subbasin group

Technical Committee on Water

Regional Prefect

Regional Adminis-trations

Prefect

Administrative Conference

Hygiene Council

Mayor

Departmental Administrations of:

Environment (DDE)

Agriculture (DDA)

Health (DDASS)

Classified Establishments Service (SCE)a

Subdepartmental Administrations

59

Its influence, therefore, is greater than appears at first sight.

The Water Distribution Companies. These are private companies which produce water for and distribute water within municipalities, having responsibility for all relevant activities from pumping through billing customers. (Only very seldom do these private companies handle sewage.) About two-thirds of the municipalities in France have negotiated long-term contracts with private water distribution companies, giving the companies the responsibility to finance and manage water distribution.

Somewhat more than half of the population is served by these companies, with the market being dominated by two, Compagnie Lyonnaise des Eaux and Compagnie General des Eaux. These two companies rank third and eighth, respectively, among service sector companies in France, and have, overall, about 50,000 employees. Their role in water management is important because, very often they act as intermediate bodies between municipalities and basin agencies, notably in questions concerning finances. For example, they bill the municipal effluent charge along with their charges for water intake provided, keep the money for two or three months, then pay it to the basin agencies. (The water distribution companies also bill water users for the sewage tax levied by the municipality. See the section on the sewage tax in chapter 7.) The companies are paid for this "administrative" service, in addition to achieving short-term earnings with the money. Consumer groups periodically raise questions about the costs of their water supply services, because there are huge disparities among municipalities in charges for water.

Actors at the River Basin Level

At the river basin level there are three main actors.

The Basin Agencies. The six basin agencies have the unique situation in France of being public establishments with an high degree of financial autonomy. Each agency is headed by a director, assisted by an executive board. The board is composed of eight civil servants and eight individuals elected by the basin committee. The primary function of the board is to approve the yearly budget of the agency.

The primary formal functions of a basin agency are to levy charges on water users, to make grants and loans to cover part of the capital costs of sewage treatment plants of municipalities and industrial activities, to make grants—in the form of premiums and superpremiums—to municipalities and industrial activities to cover part of the operation and maintenance costs of sewage treatment plants, and to develop the multi-year program of spending by the agency. Informally, staffs of basin agencies provide technical advice to all types of dischargers. Agreement on the multi-year program and on the levels of charges is requested by the basin agency from the Basin Committee (Water Parliament) and the Delegated Basin Mission.

It is important to emphasize that the basin agencies in France are not water resources (or even water quality) management agencies in the full sense of the term "management," as are the river authorities in England and Wales and, to a lesser degree, the Genossenschaften in the Federal Republic of Germany. The French river basin agencies cannot design, construct, or operate facilities;[12] they cannot set discharge standards or ambient water quality standards; they do not give permits for withdrawals or discharges; they have no police or regulatory powers. They are basically financial agencies, as their names imply, i.e., Agence Financiere de Bassin Seine-Normandie.

The Delegated Basin Mission. This body is composed of twelve civil servants from the involved ministries. Its secretary is the director of the basin agency. The Delegated Basin Mission is the link with the Interministerial Mission on Water in connection with the formulation of the multi-year programs of the basin agencies.

The Basin Committee. This controlling and advisory body consists of forty to sixty members, one-third from central government agencies, one-third from water users, and one-third elected officials--senators, deputies, mayors, members of municipal councils. The committee approves the levels of effluent charges and elects eight members of the executive board of the agency from among the members of the committee.

Actors at the Regional Level

The Technical Water Committee. This committee is attached to the prefect of the region, to coordinate water policies at the regional level. Its members are appointed from the regional staffs of the involved ministries. Its role is basically consultative.

The Regional Council. This is an advisory body, concerned primarily with allocation of funds for water management.

The Regional Administration. This body is comprised of representatives of departmental offices of the various ministries. Its role is solely coordination; it has no specific functions.

Actors at the Departmental Level

All ministries are national organizations, each with a minister and
all linked to the prime minister. Each ministry has departmental representatives (departmental administrations), each of which is linked to a departmental director and a prefect. Sometimes there are also subdepartmental
representatives, similarly linked to subprefects. Always there are regional
representatives linked to regional directors of the ministries and to the
regional prefects.

Departmental Administration of the Ministry of the Environment (DDE).
Responsibilities of the DDE include urban development planning, building
permits, public facilities planning and programming, building and maintenance of facilities such as roads and bridges. It has authority over
domainial waters and rivers in urban areas and, as such, is part of the
departmental authorization process for discharges to water bodies. The
DDE is under the dual supervision of the Minister of Environment and the
prefect. The DDE has several hundred employees per department as an
average, and also has a subdepartmental organization. Its powers and
budget make it a major actor in a department.

Departmental Administration of the Ministry of Agriculture (DDA).
The DDA is responsible for rural development planning and agricultural
questions. It has authority over non-domainial waters and, as such, is
part of the departmental authorization process for discharges to these
water bodies. The DDA is under the dual supervision of the Minister of
Agriculture and the prefect. The DDA has several hundred employees per
department as an average.

Departmental Administration of Sanitary and Social Action (DDASS).
The DDASS is part of the Ministry of Health and is responsible for

managing all sanitary and social institutions, such as hospitals, as
well as insuring that all sanitary conditions are fulfilled concerning
drinking water intakes and sewage and water treatment systems, according to the departmental health code. It is under the dual supervision
of the Minister of Health and the prefect. The DDASS has several
hundred employees per department as an average, and has an important
coordinating role in the Hygiene Council of the department, by which
authorizations for discharge are made, and in which all the other
departmental directions are represented.

The Classified Establishments Service (SCE). This agency,
which is organized at national and regional levels, is closely linked
to the Ministry of Industry. It is responsible for the authorization
procedure concerning classified establishments, which procedure
includes recommendations on plant production processes, and the
regulation of their discharges.[13] The SCE has many fewer personnel
than the departmental directions and, more than the others, keeps
close contacts with its national organization.

The Prefect (of the department). The prefect has, theoretically
and often in practice, a key role in the department, because he
issues the various prefectoral decrees for all authorizations,
including those regarding water, such as for effluent standards and
water quality objectives. He also has important police powers. He
is under the supervision of the Ministry of the Interior, but also
represents the national government. The prefectoral services are
also organized on a subdepartmental basis with two to four subprefects
representing the prefect in cities of lesser importance than the
departmental capital.

The Hygiene Council of the Department. This body is entitled
to make recommendations to the prefect with respect to wastewater
discharge authorizations. It is composed of representatives of the
various departmental directions and of water users.

The Administrative Conference. Strictly speaking, this is not
a body, but rather a procedure of internal consultations and negotiations within the departmental directions to decide upon a unified
position.

The General Council. This body is elected from among the mayors
in the department. It has both advisory and money allocation functions.

Actors at the Communal Level

The municipalities, through their elected councils and mayors,
have a consultative role in the authorization procedures. They can
also initiate the process for investment in water management facilities.

Other Actors

There is an assortment of other actors involved in water management in France, with varying degrees of influence.

Fishermens Associations. These associations are very numerous
and are quick to respond to ambient water quality conditions which
can harm fish.

Official Laboratories. These are public or private laboratories,
approved by the government with respect to such factors as equipment,
personnel, procedures. These laboratories can make analyses of
discharges and of ambient water quality, prior to judicial action.

Manufacturers Association. The manufacturers of equipment for, and components of, sewer systems and sewage treatment plants are grouped in a professional association, which participates formally and informally in water management decisions.

Judicial System. The system can be called upon to compel an activity to reduce nuisances stemming from its discharges, or to challenge a prefectoral decree. In fact, usually prior to any judicial action, there is negotiation between, for example, a fishermen's association and the plant manager, to try to decide on direct compensation. This procedure, which is done under the supervision of the departmental administration, prevents most of the cases from going to court. Because the costs involved in court action are extremely high, and because the time delays may extend for several years (whether the plaintiff's case is strong or not), it is in the plaintiff's interest to avoid going to court.

NOTES

[1] The water and riverbed of domainial waters belong to the nation.

[2] The prefect is a civil servant appointed by the national government to head a department. The appointment is a professional, not a political one, even though the prefect represents the national government in his department. The job of the prefect is to ensure application of the laws and orientations given by the national government. The prefecture is the set of institutions and services directly linked to the prefect. The prefect issues departmental decrees. He has more direct links with the Ministry of Interior than with any other ministry.

[3] The last nomenclature was established by the decree of 29 December 1977.

[4]The articles relating to detergents refer to materials whose main components are surface agents belonging to one of the following categories: anionic, cationic, ampholyte, nonionic. It is forbidden to discharge into surface, underground, and marine waters a detergent when the average biodegradibility of the surface agents which enter its composition is less than 90%, for each one of these categories.

[5]See chapter 5 and Section II.

[6]The regional prefect is, at the same time, the prefect or head of the most important department of the region.

[7]The members of the council are not directly elected by the citizens of the region. Rather, they are either elected from among the members of the municipal (communal) councils by their fellow council members, or they are elected to other offices, e.g., general council, senate, national assembly, thereby automatically becoming members of the regional council.

[8]For the Ministries of Education and Industry, the divisional unit is not the department.

[9]See Section II for explanation of premium and superpremium.

[10]This procedure evolved because municipalities refused to pay the effluent charge directly, as explained in Section II.

[11]These various bodies are described below.

[12]Sewage treatment plants are constructed and operated by municipalities, industrial activities, and a few private water companies. Reservoirs are designed, constructed, and operated by regional and departmental agencies, by public and private water distribution companies, and by Electricite de France.

[13]The SCE is part of the Service of Mines. Another activity of the Service of Mines is the regulation of ground water.

Chapter 3

ELABORATION AND IMPLEMENTATION
OF WATER POLICY AT THE NATIONAL LEVEL

THE ELEMENTS OF A
NATIONAL WATER POLICY

National water policy in France can be thought of as being comprised
of four elements: (1) regulatory laws and decrees; (2) legal interpretations, internal orientations, and prefectoral decrees relating to
regulations; (3) financial procedures; and (4) the relationship between
the national plan for socioeconomic development and water planning.

Regulatory Laws and Decrees

The first source of national water policy is obviously a general
law, such as the Law on Water of 1964. But beyond the act of preparing
and passing such legislation, there is the continuous process of promul-
gating decrees which give real substance to the law with respect to
specific applications. This process, which is fundamental, is purely
internal to the administration and reflects the evolution of the national
orientation. For example, an expansion of the scope of the decrees of
13 May 1975 represents application of regulatory power to non-classified
establishments.

Legal Interpretations, Internal Orientations, and Prefectoral Decrees

A second source of policy is comprised of the interpretations given to the laws and decrees, expressed through orientations given within the administration itself. For example, the water discharge authorization procedure can be made more or less stringent depending on prefectoral inclination. This itself partly depends on perspectives given to all prefects by the national government. Another example is the internal "orientation" of the Service of Classified Establishments, which defines "normal operations" and "normal effluents" for a given sector. This provides a margin for maneuver which can be influenced by national perspectives.

Financial Element: The Multi-Year Programs and Levels of Charges of the Basin Agencies

The basic planning tool of a basin agency is its "multi-year program" which covers five years and spells out the budgets and actions of the agency.[1] The program is prepared jointly by the director of the basin agency and the Delegated Basin Mission, in coordination with the Water Service of the Ministry of the Environment. A fair amount of the program comes from the basin agency. However, because the Delegated Basin Mission represents the various ministries interested in water management, and because the Water Service is also in direct contact with the national orientation through the Interministerial Mission on Water, it is clear that the program incorporates and expresses a national orientation.

It is important to emphasize that the formulation of the program of a basin agency leads <u>directly</u> to the delineation of the levels of abstraction charges and effluent charges in the basin for the next five years, as is described in section II. Therefore the national orientation with respect to the programs of the basin agencies also directly reflects the intent to obtain the necessary funds through the charges. Thus, the two issues--program and charges--are defined together. In this connection, the Ministry of Economics and Finances plays a role, which role can be important.

The Commission on Water of the
National Plan for Socioeconomic Development

Every five years a national plan for social and economic development is formulated. (The plan is only indicative for the private sector.) Among the many commissions, e.g., transportation, energy, housing, involved in the formulation of the plan, is the Commission on Water. It is composed mainly of the same individuals as those preparing the basin agency programs. The Commission provides another means for national orientations to be incorporated in planning and implementation. Hence, there are clear connections between the report of the Commission on Water and the multi-year basin programs. As article 14.2 of the Law of Water of 1964 states: "...the multi-year intervention program is drawn in coherence with the orientations of the plan of social and economic development."

COORDINATION OF THE
NATIONAL WATER POLICY

A national policy exists only if there is coordination among
the various groups involved, coordination which in fact achieves the
consensus and coherence which make a national policy a reality.[2]
Coordination must be insured among: (1) the various ministries,
which represent the sectoral (functional) interests; (2) the various
entities at the same geographical scale, i.e., the six basins must
be coordinated among themselves, similarly the twenty-two regions
and the ninety-one departments; (3) the various geographical
scales--the nation with the basins, with the regions, with the
departments; and (4) the basic aspects of the water management
policy--the regulatory and the financial. Finally, coordination
requires necessary articulation between the technical people and the
political people.

The Institutional Aspect of Coordination

The first aspect of coordination relates to the institutions
and commissions especially designed to achieve coordination. These
entities are shown in figure 8. Thus:

- Coordination among sectors is achieved at the national
  level among the various ministries by the Interministerial
  Mission on Water (the chairman of which is the prime
  minister); at the basin level by the Delegated Basin
  Mission; at the regional level by the regional prefect
  and the Technical Committee on Water; and at the depart-

Figure 8.   Channels of Coordination for Water Management in France

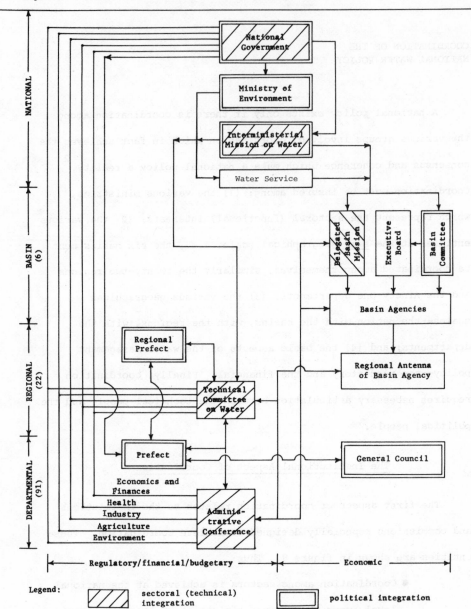

ment level by the prefect, the Hygiene Council of the
department, and the Administrative Conference.

- Coordination among governmental entities at the same
  geographical scale is achieved by regular meetings of their
  representatives, e.g., directors of basin agencies and
  prefects in the same region, and ad hoc meetings between
  the same types of representatives, e.g., between two
  directors of basin agencies, among three departmental
  or regional prefects.

- Coordination between governmental entities at two
  different geographic scales is achieved through arbitra-
  tion by representatives of the higher geographical
  scale, e.g.: Interministerial Mission for the regional
  prefects and the basin agencies; regional prefect and
  Technical Committee on Water for the regions; the prefects, Administrative Conference, and Hygiene Council
  for the communes.

- Coordination between the regulatory and financial elements
  of the national water policy is achieved by The Service
  of Water and the Interministerial Mission at the national
  level, and the Delegated Basin Mission and the Technical
  Committee on Water at the basin-region scale.

- Coordination between the technical and political aspects
  is achieved by: the Interministerial Mission on Water
  at the national level; the basin committee at the basin
  level; the prefects at regional and departmental levels.

The elected assemblies--national, regional, departmental--
obviously play a role in this articulation, but only in a
second stage.

The Human Aspect of Coordination

To achieve effective coordination is always more than a purely
institutional or organizational problem; it is also a human question.
If, in fact, a national water policy exists in France, it is not
only because institutions as indicated above have been established
to achieve coordination, but also because: (1) there is a professional
cadre, relatively small in size, involved with water problems; and
(2) this cadre is composed primarily of individuals who belong to the
three engineering corps in France. These are the Corps of Bridges
and Roads (Corps des Ponts et Chaussees), the Corps of Waters and
Forests (Corps des Eaux et Forêts), and the Corps of Mines (Corps
des Mines). Thus, the individuals who belong to the Delegated Basin
Mission, the Interministerial Mission on Water, and to the Water
Service--who are also the directors of the basin agencies and heads
of the technical committees on water in the regions--are a relatively
small group of people, homogeneous and professional, who have
personal relationships. Furthermore, and this is crucial, each of
the three corps has a special relationship with one ministry: the
Corps of Bridges and Roads with the Ministry of the Environment; the
Corps of Waters and Forests with the Ministry of Agriculture; and
the Corps of Mines with the Ministry of Industry. In addition, the
departmental directors of the ministries of Environment and Agriculture

always belong to one or the other of the first two corps, and the departmental representatives of the Ministry of Industry and of the Classified Establishments Service belong to the Corps of Mines.

Thus, coherence in national water policy is derived in part from the common traditions and the personal interrelationships established through background, training, and experience as members of the three corps.

OBJECTIVES OF THE
NATIONAL WATER POLICY

The objectives of French national water policy are expressed, explicitly and implicitly, in terms of general objectives and specific objectives. These are illustrated and discussed in this section.

General Objectives of the National Water Policy

The 1964 Law on Water calls for the satisfaction or reconciliation of the demands for water for: drinking; agriculture, industry, transport, and all human activities of general interest; biological life in water bodies, particularly fish life, as well as recreation and conservation of scenic areas; and conservation of water supplies. Interestingly, nowhere is it explicitly stated that efficiency and equitable treatment of water users are general objectives. Nevertheless, these are de facto central objectives of national water management policy. Together, the above elements represent the general objectives of national water policy in France.

Specific Objectives of the
National Water Policy

From the five water policies shown in figure 3 are derived
specific objectives. The following are some examples.[3]

Coming from the ambient water quality objectives policy, each
water course--or reach within a water course--is to be assigned a
use--or potential use--and therefore a related ambient water quality
which becomes the water quality objective to be achieved by a certain
date. Such objectives have already been defined for a certain munber
of water courses and the production of departmental maps of water
quality objectives is in progress at this time (1979). Table 2
illustrates the nature of the use designations and the associated
ambient water quality objectives, using the river Vire as an example.

Sectoral effluent reduction objectives are specified in the
so-called sector contracts. They usually stipulate a percentage
reduction of effluent discharge and a time horizon in which this
objective is to be achieved. Usually they prescribe 80% reduction
of discharge within the next five years, e.g., from pulp and paper
industry, starch manufacturing, distilleries, breweries, and beet
sugar processing plants.[4] The base from which the reduction is to
be measured is usually the level of discharge at the time the sector
contract is signed.

So-called "classical pollutants" are defined as oxidizable
material and suspended material. It was stated at the beginning
of the VIth Plan that the "desirable objective" was to reduce the
discharge of classical pollutants by 80% in 15 to 20 years; the

Table 2. Use Designations and Related Ambient Water Quality
          Standards for the Vire River

Use Designations

River Reach	Use
Source to the agglomeration of Vire	Production of potable water
From Vire to the confluence with the Souleuvre	Sport fishing
From the confluence with the Souleuvre to 1 km upstream from the confluence with the Baudre	Sport fishing and recreation
From 1 km upstream from the confluence with the Baudre to Saint-Lo	Production of potable water
From Saint-Lo to the estuary	Sport fishing and recreation

Ambient Water Quality Objectives

Location of control point	Ambient Water Quality Objectives				
	pH	% oxygen satura-tion	BOD_5 mg/l	COD mg/l	TSS mg/l
Bridge on departmental route from Sainte-Marie-Laurent to Carville	5.5-9.0	$\geq$ 70	$\leq$ 5	$\leq$ 25	$\leq$ 30
Bridge of Gourfaleur	6.5-8.5	$\geq$ 90	$\leq$ 3	$\leq$ 20	$\leq$ 30
Bridge of Saint-Fromond	5.5-9.0	$\geq$ 70	$\leq$ 5	$\leq$ 25	$\leq$ 30
Estuary at point of Gravin	5.5-9.0	$\geq$ 70	$\leq$ 5	$\leq$ 25	$\leq$ 30

Abbreviations: BOD_5, 5-day biochemical oxygen demand; COD, chemical
                oxygen demand; TSS, total suspended solids

Source: Decree No. 77-264, 16 February 1971

"minimum objective" was to keep discharges of classical pollutants to water courses at their 1970 levels. This objective implies a corollary objective in terms of sewage treatment plant construction (and operation). This corollary objective is discussed in more detail below.

With respect to toxic materials, the VIIth Plan stated that, beginning in 1975, all toxic materials generated by industrial activities shall be treated in the next ten years. This means discharge to a communal treatment facility if feasible, to a special detoxication center, or to approved on-site treatment. This objective means, in theory, that only about 10% of the toxic materials generated by industrial activities will be ultimately discharged. Note that the objective does not apply to nonpoint sources, such as urban storm runoff, a major source of heavy metals and other toxic materials, and agricultural activities, a major source of pesticides of various types, some of which are toxic.

With respect to ground water protection, intake of ground water is discouraged when it is for industrial purposes which do not require water of high quality. Discharge of wastewater to ground water aquifers is usually forbidden.

THE OBJECTIVE OF "ELIMINATION" OF
CLASSICAL POLLUTANTS BY 80%

This objective has special importance because it is the one which has drawn the most attention and financial effort thus far, and will continue to do so for at least the next 10 years, i.e., through the decade of the eighties. This is why it is appropriate to present further details about this objective.

Table 3 shows treatment plant capacity in inhabitant-equivalents constructed in the five-year period from the beginning of 1971 to the end of 1975 in each of the basin agency jurisdictions, subdivided into municipal plants and plants of industrial activities not connected to municipal sewage systems. Table 4 shows, for the same categories and the same units (inhabitant-equivalents), the implications of the objective of "elimination" of 80% of classical pollutants. The table shows: 1970 generation; 1970 discharge, with the estimated average 60% removal efficiency; 1990 estimated generation; 1990 desirable and minimum discharge objectives; and the related 1990 desirable and minimum installed treatment plant capacity objectives, assuming 90% removal efficiency. An inhabitant-equivalent is defined in France as consisting of a daily discharge of 90 grams of suspended material and 57 grams of oxidizable material (OM). The oxidizable material is measured by a combination of chemical oxygen demand (COD) and five-day biochemical oxygen demand (BOD_5), according to the following formula:

$$OM = \frac{COD + 2\ BOD_5}{3}$$

The sources of the estimated 1990 municipal treatment plant influent are shown in table 5. The result corresponds to an annual increase of 4.1% in the quantity of influent to municipal treatment plants between 1970 and 1990. A major part of this increase is due

Table 3. Treatment Plant Capacity Built Between 1970 and 1976[a]

Basin Agency[b]	Treatment plant capacity existing in 1970			Treatment plant capacity existing or being built on 1 January 1976			Treatment plant capacity built or being built in 5-year period		
	M	I	T	M	I	T	M	I	T
AG	1.1	1.0	2.1	3.6	2.9	6.5	2.5	1.9	4.4
AP	1.3	4.3	5.6	3.8	10.3	14.1	2.5	6.0	8.5
LB	2.4	5.0	7.4	7.3	7.0	14.3	4.9	2.0	6.9
RM	0.6	4.7	5.3	2.9	11.0	13.9	2.3	6.3	8.6
RMC	2.5	0.6	3.1	8.0	11.0	19.0	5.5	10.4	15.9
SN	6.0	3.9	9.9	16.0	9.5	25.5	10.0	5.6	15.6
Totals	13.9	19.5	33.4	41.6	51.7	93.3	27.7	32.2	59.9

[a]All units are 10^6 inhabitant-equivalents (I.E.). One I.E. = a daily discharge of 90 grams of suspended matter plus 57 grams of oxidizable material.

[b]Basin agencies are: AG, Adour-Garonne; AP, Artois-Picardie; LB, Loire-Bretagne; RM, Rhin-Meuse; RMC, Rhone-Mediterranee-Corse; and SN, Seine-Normandie.

Abbreviations: M, municipalities, including industrial activities discharging to municipal sewage systems; I, industrial activities discharging directly, i.e., not connected to municipal sewage systems; and T, total, i.e., M + I.

Sources: Computed from data in: Fournier, Y. and Philip, R. P., 1978, Assainissement: evolution, et developpement, Nuisances et Environnement, No. 71, June; Environnement et cadre de vie: dossier statistique, Tome 2, 1978, Ministry of Environment and Quality of Life, Paris.

Table 4. Estimated Installed Treatment Plant Capacity Required to Meet Desirable and Minimum Objectives for 1990[a]

| Basin, Agency[b] | 1970 Generation | | | 1970 Discharge | | | 1990 Estimated Generation | | | 1990 Discharge Objectives | | | | | | 1990 Installed Treatment Plant Capacity Objectives Assuming 90% Removal Efficiency | | | | | |
| | | | | | | | | | | Desirable Objective[d] | | | Minimum Objective[e] | | | Desirable Objective[d] | | | Minimum Objective[e] | | |
	M	I	T	M	I	T	M	I	T	M	I	T	M	I	T	M	I	T	M	I	T
AG	4.2	4.4	8.6	3.5	3.8	7.3	10.0	8.8	18.8	2.0	1.8	3.8	3.5	3.8	7.3	9.6	7.5	17.1	7.3	5.5	12.8
AP	6.4	8.6	15.0	5.6	6.0	11.6	14.4	18.8	33.2	2.9	3.8	6.7	5.6	6.0	11.6	12.0	17.3	29.3	9.7	14.2	23.9
LB	7.4	6.8	14.2	6.0	3.8	9.8	13.8	14.4	28.2	2.8	2.9	5.7	6.0	3.8	9.8	13.8	14.0	27.8	8.6	10.6	19.2
RM	4.1	10.6	14.7	3.7	7.8	11.5	8.1	23.1	31.2	1.6	4.6	6.2	3.7	7.8	11.5	7.9	20.8	28.7	4.9	17.0	21.9
RMC	10.7	12.1	22.8	9.2	11.7	20.9	21.3	25.0	46.3	4.3	5.0	9.3	9.2	11.7	20.9	18.5	23.0	41.5	13.4	14.8	28.2
SN	17.9	13.9	31.8	14.3	11.6	25.9	45.0	35.0	80.0	9.0	7.0	16.0	14.3	11.6	25.9	38.5	30.5	69.0	34.0	26.0	60.0
Totals	50.7	56.4	107.1	42.3	44.7	87.0	112.5	125.1	237.7	22.5	25.0	47.5	42.3	44.7	87.0	100.0	113.1	213.1	77.9	88.1	166.0

[a]All units are in inhabitant-equivalents (I.E.). One I.E. = a daily discharge of 90 grams of suspended matter plus 57 grams of oxidizable material.

[b]Basin agencies are: AG, Adour-Garonne; AP, Artois-Picardie; LB, Loire-Bretagne; RM, Rhin-Meuse; RMC, Rhone-Mediterranee-Corse; and SN, Seine-Normandie.

[c]Based on: (1) assumed proportion discharged to treatment plants equal to treatment plant capacity specified in table 3; and (2) assumed 60% removal in treatment.

[d]Desirable objective: discharge be equal to or less than 20% of generation.

[e]Minimum objective: discharge be no larger in 1990 than in 1970.

Abbreviations: M, municipalities, including industrial activities discharging to municipal sewage systems; I, industrial activities discharging directly, i.e., not connected to municipal sewage systems; T, total, i.e., M + I.

Sources: Computed from data in: Fournier, Y. and Philip, R. P., 1978, Assainissement: evolution, et developpement, Nuisances et Environnement, No. 71, June; Environnement et cadre de vie: dossier statistique, Tome 2, 1978, Ministry of Environment and Quality of Life, Paris.

to the estimated increase in discharges from industrial activities
into municipal sewage systems. The estimated 1990 industrial influent
for activities not connected to municipal systems was computed by
using the same annual increase of 4.1%, yielding 125.1 million
inhabitant-equivalents. To reach the 1990 desirable objectives
in relation to reduction in discharge of classical pollutants,
installed treatment plant capacities within the jurisdictions of the
basin agencies would have to be multiplied between 1976 and 1990 by
the following factors: Adour-Garonne, 2.63; Artois-Picardie, 2.08;
Loire-Bretagne, 1.94; Rhin-Meuse, 2.06; Rhone-Mediterranee-Corse,
2.18; and Seine-Normandie, 2.71.

Table 5. Sources of Estimated 1990 Municipal Treatment
Plant Influent

Source	Influent, 10^6 I.E.
Resident population	59.0
Seasonal population	15.5
Other domestic uses of water	7.9
Industrial activities discharging into municipal sewage systems	30.1
Total	112.5

I.E. = inhabitant-equivalent

The figures in the preceding paragraphs of this section provide
some perspective on past performance and future "desired" performance.
Between 1970 and 1976, treatment plant capacity almost tripled.

For the desirable treatment plant capacity objective for classical
pollutants to be achieved by 1990, the annual growth rate of installed
capacity should be about 6.6% between 1976 and 1990. Because the
facilities built tend to be, on the average, smaller when more and
more municipalities and industrial activities are equipped, the
costs to build these facilities would be expected to increase more
than 6.6% per year.[5] Note that in connection with both the desirable
and the minimum objectives, treatment efficiency is assumed to
increase from the 60% average achieved in the 1970-1976 period,
to 90% in 1990. This means that an integral part of the objective
relates to the design and operation of the treatment plants, i.e.,
removal efficiency actually achieved.

Finally, it should be emphasized that the 1990 estimates of
treatment plant influent and discharge for the area of each basin
agency are crude. They do not explicitly take into account such
factors as:

(1) source reduction of generation and discharge through
recycling, production process changes, materials and
energy recovery, and better management;

(2) the evolution of the national and regional industrial
mixes and changes in the regional locations of
activities;

(3) the possible or probable change in the value of the
"inhabitant-equivalent," which was assumed constant
at its 1970 value; and

(4) contributions from nonpoint sources.

Even so, these figures, or figures like them, are used in developing

the multi-year programs of the basin agencies. To provide a better

basis for the formulation of such programs requires an analysis

which would look at the implications of alternative combinations of

these factors.

NOTES

[1] For more details, see chapter 5 and section II.

[2] Practically, much of the coordination is achieved in France
through a peer group of civil servants from the various corps, as
noted below.

[3] In addition to the specific objectives illustrated with
respect to water quality management, there are also specific basin
objectives with respect to flood damage reduction and irrigation,
where relevant.

[4] The objectives expressed in the sector contracts are national
objectives. They do not apply to specific sources, and they have no
legal standing. They are used as background by the Interministerial
Mission and the basin agencies in the formulation of the multi-year
basin programs.

[5] However, since 1976 these increased costs have not occurred,
because of reasons explained in section II.

Chapter 4

AUTHORIZATION PROCEDURES WITH RESPECT
TO WASTEWATER DISCHARGES AND WATER INTAKES

GENERAL PERSPECTIVE ON THE AUTHORIZATION
OF WASTEWATER DISCHARGES AND INTAKES

There are two main characteristics of the authorization process.
First, what is stated in the decrees are the procedures for obtaining an
authorization and the limits of size below which the procedure need not be
applied. In fact, there are practically no nationally defined standards,
with respect to either effluents or ambient water quality. Second, a
large, and to a certain extent discretionary, power is granted to the
prefect. He decides what standards are to be applied to a specific discharger and to a specific water course, and makes the final decision with
respect to issuing the permit (authorization). Even cases involving
activities which are classified establishments, which lead to the intervention of the Classified Establishments Service in the authorization
process, and the recent environmental impact statement procedure, which
provides for the involvement of the public, do not change the fact that
the authorization is basically what the prefect wants it to be.

To assess the practical role and use of the authorization procedures,
the question is therefore to understand the system of influence and
power in which the prefect acts and issues permits. This system

varies from one department to another and even from case to case

within the same department. Nevertheless, one finds always the

following actors, in addition to the prefect: the discharger,

municipality or industrial activity, and the various departmental

services, usually Agriculture (DDA) or Environment (DDE), and sometimes

the Classified Establishments Service (SCE). In some rare cases

in which conflict exists, the public also becomes involved. Usually

there is a real balance of power between the prefect and the DDA or

DDE. This means that the authorization decisions usually need an

administrative consensus and cannot be imposed by one side only.

In addition to this general characterization of the authorization

process, three other factors affect the process. One, the DDA and

DDE have many other, more important, responsibilities beside discharge/

intake authorizations. They are responsible for all rural and urban

public works in which they are financially interested. That is, these

departments get a percentage of the total value of the work which is

constructed under their responsibilities, including all municipal

works. Very often, there is implicit negotiation between the discharger

and the departmental services about the various points which they have

to settle, the intake/discharge authorization being only one of them.

This results in a sort of package deal. Two, for existing discharges

and intakes, there is a grossly insufficient number of people to

review the measures and so-called compulsory declarations to be made

or authorizations requested. Three, the economic situation in the

period 1974 to 1978 was extremely difficult. Industry successfully used

the unemployment argument to widen its margin of maneuver regarding

conditions for authorizations.

Despite these shortcomings, the authorization procedures are useful for four reasons. First, they insure that new establishments do not affect the water system much more than an average establishment of its type and size. Second, they give the possibility for the prefect and/or the public to react and demand improvements after a pollution accident or significant degradation of water quality has occurred. Third, they set the ground for future improvements when the general situation or the public concern will make them possible. Fourth, they allow enquiry and public participation for major water projects, such as dams and nuclear power plants.

There are two basic sources of the authorization procedures, the Law on Water of 1964 and the Law on Classified Establishments of 21 September 1977. The Law on Water states: "...decrees of the Council of State shall determine the conditions by which can be regulated or forbidden...discharges and all factors which can alter superficial or underground water quality" (article 6); and "...the owners of facilities which discharge to the waters shall have to take all dispositions to satisfy the conditions which will be imposed on their discharges...so that the ambient water quality reaches the required characteristics" (article 4). Applicable decrees stemming from this law are: the 23 February 1973 decree which delineates the authorization process itself; the 12 March 1975 decree which defines how a measurement to control a discharge should be done; and the 3 decrees of 13 May 1975 which define classes of discharge quality for municipal treatment plants, set the thresholds under which no authorization is needed, and set discharges about which the consultation of regional or national bodies is required.

The law on Classified Establishments defines two classes of establishments (activities) among the classified establishments: those which must have an authorization before starting production; and those which have only to make a declaration of the location of intake and/or discharge. The authorization process involves a public enquiry, the preparation of an environmental impact statement (EIS), and can involve specific in-plant modifications required at the national level or specifically for one establishment.

The major decree with respect to the authorization procedure is that of 21 September 1977, which establishes the procedure and delineates some specific points which the EIS must address. The nomenclature of classified establishments is updated by decree.

CONDITIONS AND PROCESS
OF AUTHORIZATION PROCEDURES

Both certain water intakes and wastewater discharges must be authorized. Because the emphasis in this report is on water quality management, only a few comments are presented with respect to water intake. Primary attention in this chapter is given to wastewater discharges.

Water Intake

Surface water intake in domainial (navigable) waters must be authorized, as well as the subsequent use of the public water channel for the intake structure. For non-domainial waters, the authorization is needed only if the flow of water is modified. These authorizations by the prefect include a public enquiry.

Underground water intake, when it is for non-domestic use and
when it exceeds eight cubic meters per hour, must be declared for
official recording. The total monthly intake must also be declared.
For some regions, including the Paris region, all underground water
intake must be authorized because there are risks of overuse. Priority
is given to withdrawals for municipal drinking use.

Every water intake for drinking use needs a "Public Utility
Declaration." This involves a public enquiry, and the determination
of a protected area to prevent the source of the drinking water from
pollution.

Wastewater Discharge

With only minor exceptions, authorization of discharges is
required for all types of water bodies--surface streams, unconfined
and confined ground water aquifers, lakes, marshes, canals, and the
sea. The decree of May 1975 defines the thresholds below which no
authorization is needed. These thresholds depend on: (1) where the
discharge is to be made, e.g., surface stream, canal, sea, underground; (2) the water quality objectives for the water body; and
(3) the magnitude of the discharge flow, e.g., liters per second.
For example, the threshold is 500 inhabitant-equivalents for discharge
into a river which has average to low water quality objectives.

An authorization must be obtained for dischargers who were
operating before the date of the decree, unless they had discharge
authorization under some other legislation.

THE AUTHORIZATION PROCEDURE FOR
WASTEWATER DISCHARGES

The procedure for obtaining an authorization by an industrial
plant for a wastewater discharge is diagrammed in figure 9 and is
discussed in the following paragraphs.[1] The numbers refer to the
indicated boxes in the figure.

(1) Contents of the application for authorization. The
application must describe: the nature of the activity; the discharge,
e.g., nature, flow, time pattern; expected changes in ambient water
quality resulting from the proposed discharge; the proposed actions
to minimize polluting discharges; and the technical solutions to be
adopted. For classified establishments and for treatment plants of
a size of more than 10,000 inhabitant-equivalents, canals, and other
major works involving water use, and more generally for all works of
a value of more than 6 million francs,[2] an environmental impact
statement must be included. The EIS must include analysis of: the
initial state of the site; the effects of the projected facility on
the environment; and the actions to be taken to reduce or compensate
for the adverse consequences of the project.

The application is then sent to the prefect for decision. For
this purpose, he has to consider the advice of the departmental
Hygiene Council, in which all departmental directions are represented.
In cases where the discharge is especially important, he will also
have to take advice from the Delegated Basin Mission, the Superior
Council on Classified Establishments, and the Superior Council of
Hygiene of France.[3] These bodies must refer the request to the

Minister of the Environment, if they do not agree on the advice to give to the prefect. Prior to the meeting of the departmental Hygiene Council for advising the prefect, two sets of events must have taken place the results of which will feed and enlighten this departmental council: the administrative set and the public enquiry set.

(2) Field inspection and administrative conference. The prefect, once he has received the file from the applicant, names the departmental direction which will be responsible for the case. For urban and domainial water of less than 10 meters depth, it is the Departmental Direction of Environment. For rural and non-domainial water of less than 10 meters depth, it is the Departmental Direction of Agriculture. For underground water, it is the Service of Mines. In the case of a classified establishment, the Classified Establishments Service (Ministry of Industry) is responsible.

Then, an engineer named by the responsible departmental service makes a visit to the site with the applicant and the concerned mayors. He then makes a report detailing the technical conditions which should be required of the applicant. This report and the advice it contains are sent to the prefect. If underground aquifers are involved, the Official Geologist will also make a report and recommendations.

Finally, an administrative conference is held, where all departmental services concerned with the discharge, directly or indirectly, are represented, so that coordination can be achieved before the meeting of the departmental Hygiene Council is held.

(3) The public enquiry. When the prefect receives the filing from the applicant, the prefect defines the date and duration of the

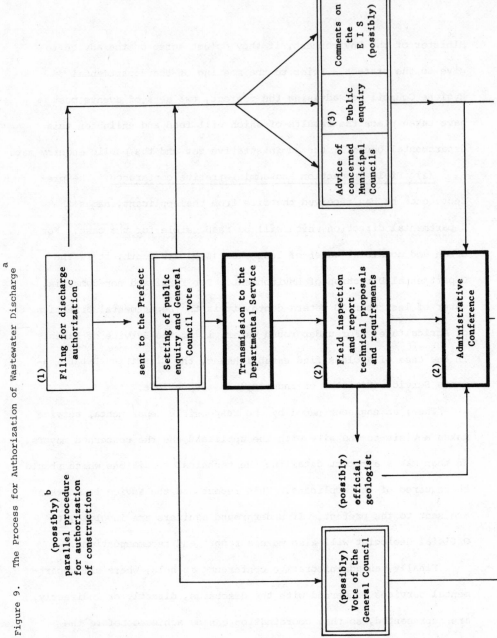

Figure 9. The Process for Authorization of Wastewater Discharge[a]

(possibly)[b]
parallel procedure
for authorization
of construction

(1)
Filing for discharge
authorization[c]

sent to the Prefect

Setting of public
enquiry and General
Council vote

Transmission to the
Departmental Service

(2)
Field inspection
and report :
technical proposals
and requirements

(possibly)
official
geologist

(possibly)
Vote of the
General Council

(2)
Administrative
Conference

(3)
Advice of
concerned
Municipal
Councils

Public
enquiry

Comments on
the
E I S
(possibly)

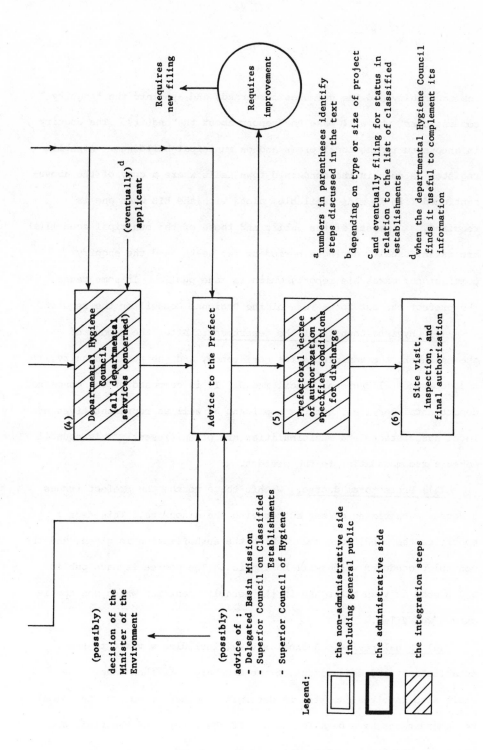

Requires new filing

Requires improvement

(eventually)[d]
applicant

(4) Departmental Hygiene Council
(all departmental services concerned)

Advice to the Prefect

(5) Prefectoral decree of authorization, specifies conditions for discharge

(6) Site visit, inspection, and final authorization

(possibly)
decision of the Minister of the Environment

(possibly)
advice of :
- Delegated Basin Mission
- Superior Council on Classified Establishments
- Superior Council of Hygiene

[a] numbers in parentheses identify steps discussed in the text

[b] depending on type or size of project

[c] and eventually filing for status in relation to the list of classified establishments

[d] when the departmental Hygiene Council finds it useful to complement its information

Legend:

☐ the non-administrative side including general public

■ the administrative side

▨ the integration steps

93

public enquiry and the communes concerned, and appoints the "enquiry commissioner" who will make the report about the enquiry. The enquiry is announced in the local press and on municipal buildings. A public register is open in the concerned town halls where a copy of the above-mentioned file is made available, along with the EIS when one is required.[4] Comments of the public and those of the municipal council(s) are taken, usually during a period of two weeks, and the enquiry commissioner makes his report, which is made public. In some cases, the prefect can ask for a vote of the Regional Council on the request.

(4) <u>Hygiene Council of the department.</u> After the report of the engineer, the administrative conference, and the public enquiry, the departmental Hygiene Council meets. It is composed of the concerned departmental services, but the applicant as well as representatives of local associations and municipalities are often present. This council makes a recommendation to the prefect.

(5) <u>Prefectoral decree.</u> Within three months the prefect issues a decree authorizing or not authorizing the discharge. This decree specifies the conditions under which the authorization is given, and the control procedures which will be applied. The decree is made public and posted in town halls and in the establishment for which the application was filed.

(6) <u>Final review.</u> A final review, including a visit to the establishment and measurement of the discharge, is then made. The visit also involves checking to determine whether or not the technical recommendations have been followed. If the results of the visit are positive, the final authorization is then issued.

THE CASE OF CLASSIFIED
ESTABLISHMENTS

For classified establishments, especially those in Category A,
the following elements comprise the authorization procedure. Fist,
the EIS is always required. Second, the application for authorization
is normally quite detailed, including specifications of in-plant
processes--which are not made public in the public enquiry if a secret
process is involved. Third, the concern of the administration is
not limited to the discharge itself, but includes elements such as
risks of accidents and work safety. This broader focus can result in
recommendations concerning not only the discharge/treatment system
itself, but also in-plant processes and procedures. Fourth, there
is a sort of double administrative process for the authorization,
consisting of: (a) the visit by the representative of the Classified
Establishments Service and the resulting recommendations and
standards; and (b) the departmental services views and wishes. Fifth,
the public enquiry lasts longer than for an activity which is not a
classified establishment, is more complete, and is widely publicized.
Sixth, visits and controls, even though fairly rare, are likely to be
more regular and effective than for non-classified establishments.

STANDARDS AND CRITERIA;
CONTROLS AND PENALTIES

Having described the authorization process, the next questions
are: (1) on the basis of what criteria are the authorizations given or
denied, and the standards in the authorizations established; and

(2) what controls are in fact applied to dischargers in order to
insure they meet the conditions of their authorization and what
penalties are imposed if they fail to meet the authorization conditions.

Standards and Criteria

Four points can be made about the standards and criteria relating
to authorizing discharges. First, discharge standards are defined
case by case taking into account: (1) the actual level of ambient
water quality of the water body; (2) its self-purification capacity;
and (3) the expected use of the water body, and hence the related
ambient water quality objectives. Second, discharges from all point
sources to surface waters everywhere must meet the following minimum
standards: (1) temperature of effluent must not exceed $30^{\circ}C$; (2) pH of
effluent must be between 5.5 and 8.5; (3) color of effluent must not
induce color in the receiving water; (4) effluent must not result in
mortality of fish beyond 50 meters downstream from its discharge point;
and (5) effluent must not have any algae or floating objects. Additional
specifications beyond these limits can be imposed on nuclear power
plants and offshore oil extraction activities. Third, all dischargers
must be equipped in such a way that flow measurements and samples of
discharge can easily be taken. Fourth, the decrees of 10 June 1976
specify   standards which an effluent from an industrial activity must
meet to be accepted in a municipal sewage system. However, it may be
possible to negotiate case by case with the municipality, depending on
the capacity of the existing municipal facility.

For non-municipal discharges, the decree of 13 May 1975 states
only that "the minimum quality of the discharge should be set for all
parameters characterizing the discharge, taking into account the
activity which is at the origin of the discharge. This minimum
quality is determined according to the conditions of use of the
receiving water, the level of pollution in (ambient quality of)
the receiving water, and the self-purification capacity of the receiving
water." For municipal discharges, the standards imposed--based on
the same decree--are chosen from among six different levels of treatment ranging from simple settling to full tertiary treatment. There
is a provision for "absence of prescription of a minimum effluent
standard," but this should be "strictly restricted to discharge in
rivers with important permanent flow." Furthermore, for discharges
into a canal or lake, "the levels I and II (partial treatment) should
only be accepted in a transitory way and for a fixed period of time,
specified in the prefectoral authorization decree." Although these
standards are expressed in terms of a level or degree of treatment,
each level _implies_ a quality of effluent in terms of _concentrations_
of certain water quality indicators.

The decree of 6 June 1953 requires that, for classified establishments discharging into rivers, total suspended solids and BOD_5 should
not exceed 500 mg per liter, and nitrogen 150 mg per liter, as (N).
If the river is fairly unpolluted, the numbers become 100 mg per liter,
200 mg per liter, and 60 mg per liter, for total suspended solids, BOD_5,
and nitrogen, respectively. For each type of activity included in
the classified establishments, recommended production and water

utilization processes are defined, as well as the corresponding
"normal" effluent discharges, which are expressed in terms of, for
example, kilograms of BOD_5 (5-day biochemical oxygen demand) and
TSS (total suspended solids), per ton of product. There are in fact
effluent and process standards which guide the Service of Classified
Establishments in its recommendations about discharge authorizations,[5]
which, according to the decree of 21 September 1977, "must take into
account the efficiency and cost of available technologies, and the
quality and use of the receiving water." The standards guide the
Classified Establishments Service in its inspection visits and can
be the basis for an action on the part of the prefect for a new
authorization procedure to be initiated.[6]

Finally, there are standards derived from the "sector contracts"
negotiated at the national level between a sector and the Ministry of
the Environment. Some of these contracts provide special financial
help, others do not. These sectors are cement manufacturing, metal
finishing, beet sugar processing plants, four food processing subsectors,
petroleum refining, and pulp for paper production.

In effect, the whole discharge standards question is related to
the ambient water quality objectives policy. This means that
standards are set, given that: (1) detailed and approved maps of
water quality objectives exist; (2) the relationships between potential
discharges and ambient water quality can be estimated with reasonable
accuracy; and (3) a rule exists, accepted by all parties, concerning
the relative effort water users have to make on the same water course
from upstream to downstream.

Controls and Penalties

There are basically two channels by which administrative and legal procedures can take place in relation to discharges: (1) by the inspector of the Classified Establishments Service; and (2) by a complaint, e.g., made by a fisherman's association.[7] Activities which are classified establishments are subject to control by a special corps of inspectors. These inspectors have a permanent right of visiting and taking discharge samples in a plant, without prior notification. They determine whether a plant is respecting the prefectoral decree authorizing the discharge. In case they want to take action, they have the prefect require the discharger to meet the requirements within a certain period of time. If this is not done, the prefect may order temporary closing of the plant until the requirements are met. However, this can involve a long and complex procedure.

In fact, things usually work differently. The inspectors, who potentially have a lot of power, use it rather as a bargaining tool to push and negotiate directly the needed improvements. Furthermore, because many of the old authorizations are not too stringent, and because it is fairly difficult to change them (a decree of the Council of State is needed case by case), the inspectors try to induce the industrial discharger to go beyond his authorization even though the inspectors have, strictly speaking, not much legal ground to require anything if the conditions set out in the authorization are being met. The only exception is when there are specific discharge standards for the sector to which the plant belongs. However, because the inspectors are not numerous enough, they only cover the most important establishments in their respective areas.

For all other cases--that is, non-classified establishments and minor classified establishments--action is only taken when there is a complaint. If it is a classified establishment against which a complaint has been made, the local inspector makes an inspection, as in the case described just above. However, in all cases where a complaint has been made, an ambient water quality sample and a discharge sample are taken and analyzed. Where these indicate that requirements have not been, or are not being met, the complainant starts judicial action. Such action is often based on: (1) nonadherence to authorization conditions; (2) nonadherence to discharge standards for the sector; and (3) infraction of article 434-1 of the Rural Code stipulating that no harm should result to fish life, nutrition, and reproduction, nor to fish nutritive value.

However, even where the authorized conditions are not being met, no one really has an interest in going to court. Typically, judicial actions involve: long and costly procedures; almost no money for the plaintiff to be compensated for his damage if he is in the right; and a lot of trouble in the future for the plant because it will have a "judicial record." Normally, the departmental service, DDE or DDA, tries to negotiate a bargain between the two parties, in which compensation for damages and an adopted schedule for improvement of discharge are traded for foregoing judicial action.

NOTES

[1]The procedure as diagrammed and discussed is also valid for a municipal discharge.

[2]About 1.5 million dollars at the 1978 exchange rate.

[3]These are national level bodies, whose members are professionals, presumably wise, and are appointed.

[4]For an activity which is a classified establishment, the public enquiry is termed a "commodo-et-incommodo" enquiry.

[5]The discharge standards established by the Service of Classified Establishments for each sector have no legal standing and do not take into account the ambient water quality objectives. For any given discharger, belonging to some sector, the standards may be too stringent or not stringent enough for a given location.

[6]See chapter 6.

[7]The national territory is literally covered by such associations.

Chapter 5

THE BASIN AGENCIES

SETTING AND ORGANIZATION
OF THE BASIN AGENCIES SYSTEM

Objectives and Functions of the Basin Agencies

Created by the decree of 14 September 1966, the six "financial
basin agencies," usually called "basin agencies" or simply "agencies,"
comprise a key element of the water quality management system in France.
They are public administrative establishments (governmental agencies),
with many attributes of a private company and a certain degree of financial
autonomy, the latter in the sense that the agencies have substantial
discretion in deciding yearly expenditures.[1] They also have a technical
role in water management. According to the decree, the objective of the
agencies is "to facilitate the various actions of common interest for
the basin...in order to insure equilibrium between water demand and
supply, to reach the water quality objectives...to improve and increase
the water resources and to provide flood control."

To achieve these objectives, the same decree states that each
agency must elaborate a "multi-year program of intervention," approved
by the Prime Minister, and that the global amount of charges to be
imposed, and received, by the agency is fixed by the cost of the program

to the agency. These programs, prepared for five years, are the
cornerstones of the actions of the agencies and the bases for setting
the charges. The technical function of a basin agency is to prepare
the multi-year program, and also to advise about the planning, development, and operation of the projects of which the program consists.
The financial function of an agency consists in levying the charges
and granting subsidies and loans to achieve its program. It should
be emphasized that the agencies have no regulatory power and cannot
commission, construct, or operate projects.

Functioning of the Basin Agency System

The basin agency system involves various interrelated elements,
linking national, basin, sectoral, and local interests. These
elements are depicted in figure 10 and are described briefly below.

Each basin agency has a director, named by the prime minister.
Because there are six agencies, each one of the three corps of
engineers "has" two agencies. Thus, the Corps of Waters and Forests
(Ministry of Agriculture) has the Loire-Bretagne and Adour-Garonne
basin agencies, the Corps of Bridges and Roads (Ministry of the
Environment) has the Artois-Picardie and Rhin-Meuse basin agencies,
and the Corps of Mines (Ministry of Industry) has the Seine-Normandie
and Rhone-Mediterranee-Corse basin agencies.

The director is in direct relationship to the executive board
of the basin agency. This board has twenty members: ten civil
servants representing the various ministries and chosen from among
the members of the Delegated Basin Mission; and ten representatives

Figure 10. Overall Organization of Basin Agency System

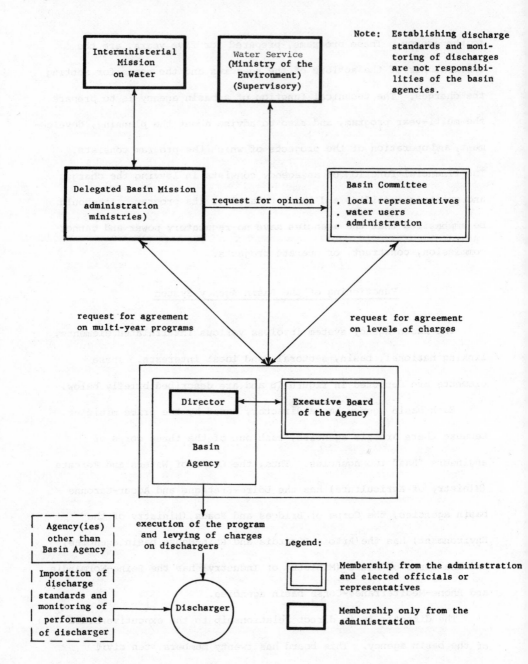

104

from municipalities and water users, elected from among the members of
the Basin Committee. The executive board controls the actions of the
director and votes on all the agency's programs.

The Basin Committee has from forty to sixty members, depending
on the agency. It is the "water parliament." It is composed equally
of: (1) representatives of the administration (ministries), who
are civil servants; (2) representatives of municipal councils, elected
by the general councils of the departments whose territories are
encompassed by the agency; and (3) representatives of the various
categories of users--agriculture, commerce, industry, National Union
of Tourism, Electricite de France, water distribution companies,
Union of Riparian Associations, fishermens associations. The Basin
Committee is systematically consulted and, in fact, has power to
affect all options regarding the policies of the agency. More
specifically, the Basin Committee is the one which votes on the levels
of the charges, which means that it also votes on the corresponding
multi-year program. The Basin Committee organizes itself in committees
which review the loans and grants proposed to be made by the basin
agency. It is kept regularly informed by the staff of the agency of
the evolution of the multi-year program. It should be emphasized
that in the Basin Committee, which has real power, civil servants
comprise only one-third of the membership.

The Delegated Basin Mission is composed of 12 civil servants
from the concerned ministries. The director of the basin agency acts
ex officio as its secretary. The Delegated Basin Mission has the role
of coordination with the Interministerial Mission on Water. It is

the interface between: (1) the agency and national policy; (2) the agency and the ministries; and (3) the agency and the regional prefects. The Delegated Basin Mission votes on the multi-year program of the agency, coordinates the programs of the national water quality objectives policy, and is consulted on authorizations for major discharges.

The Water Service of the Ministry of the Environment is responsible for supervision of the basin agencies. Theoretically this means the final decision power. In fact, this supervision has been extremely light, because the inclusion of the national orientations in the agencies' programs is usually made without the need of further coordination at the level of the Service of Water.

Finally, as noted in chapter three, interagency coordination and the national-intersectoral coordination are achieved, beyond the formal setting which has been presented, through the cohesiveness of a relatively small group of public servants. These public servants represent their ministries in the Interministerial Mission on Water, the executive boards of the agencies, the basin committees, and in the delegated basin missions. The members of the Water Service of the Ministry of the Environment and the directors of the agencies belong naturally to this small group, which is all the more homogeneous because all its members belong to the three corps of engineers of the administration.[2]

THE ACTIVITIES
OF THE BASIN AGENCIES

In the period 1979 through 1991, the agencies together will levy
charges in an amount averaging a little less than half a billion
dollars per year. They will therefore grant subsidies and premiums
of exactly that sum of money, minus the agencies' operating costs,
for all kinds of projects. Because the agencies do not construct
nor commission anything themselves, their action is basically financial.
It consists of: (1) inducing the actors--national government, departments, regions, municipal councils, and the private sector--through
grants and loans to initiate projects which are consistent with the
agencies' programs; and (2) giving premiums--and in some cases also
superpremiums--to those activities which reduce their discharges.
The question then is: what types of projects and activities are
subsidized, according to what criteria, and by what kind of procedure?

What is Subsidized and under What Conditions

Table 6 lists the types of projects and activities which are
subsidized. Not all basin agencies subsidize all the types listed;
all the types are subsidized by at least one basin agency.

Grants by the basin agencies range from 15% to 50% of the capital
cost, and loans (with a low interest rate) range from 20% to 50% of
the capital cost. The percentage depends on the kind of project,
who is the beneficiary (e.g., industrial activity, municipality,
or national government), and the zone in which the project

Table 6.   Types of Projects and Activities Subsidized by
            Basin Agencies

PROJECTS AND ACTIVITIES RELATED TO WATER QUALITY MANAGEMENT

Capital costs of municipal facilities

      Sewage treatment plants, including tertiary treatment
      Modifications of existing treatment plants
      Intercommunal collectors (interceptor sewers)
      Sewer systems (collection pipes and related facilities)
      Discharge pipes (outfalls) in the sea

Capital costs of facilities of industrial activities

      Source reduction and treatment facilities for organic materials
      Source reduction and treatment facilities for toxic materials
        and salts
      Facilities for handling/modifying solid or toxic residuals
        which otherwise would have adverse effects on water quality

Other

      Elimination of solid residuals repositories near surface waters
        or aquifers
      Facilities to treat petroleum residuals
      New treatment technologies
      Premium for reduction in discharges, any type of activity
      Superpremium for good efficiency of treatment plant operation,
        any type of activity
      Departmental technical assistance

PROJECTS AND ACTIVITIES RELATED TO WATER QUANTITY MANAGEMENT

      Dams and reservoirs
      Interconnections between and among water supply systems (for
        transfer of water)
      Measures to protect existing water intakes
      New water intake facilities
      Measures for managing water demand within industrial activities
      Transfers of intakes from underground sources to surface water
        sources
      Measures for protecting aquifers against pollution accidents
      Reservation of land
      Measures to reduce infiltration of water into water distribution
        and sewage networks

is to be implemented.[3] For discharge reduction facilities, there is a limit to the cost of reduction per unit which is acceptable for any given pollutant.[4] This limit also depends on the zone, and on the size of the facility.

In addition to the subsidies by the basin agencies, the national government (or regions or departments) give subsidies (grants) to public entities for dams, water transfers, and sewage collection and water distribution networks. From these sources the only subsidies to industrial activities are those in the sector contracts and for experimental or new less-polluting technologies.

To illustrate, for the third multi-year program (1977-1981), subsidies from the Seine-Normandie Basin Agency have been allocated as shown in table 7.

CURRENT AND FUTURE PROBLEMS
AND PROGRAMS OF THE BASIN AGENCIES

The problems cited below have already been recognized by the basin agencies, and usually policies have been initiated to resolve them. Nevertheless at this date, 1979, they still represent problems. In some cases no satisfactory action has yet been designed and much time may be needed to achieve satisfactory results. The order of listing of the problems does not imply order of importance or priority.

● Pollution by nitrates, especially in the ammonium ion form (NH_4^+), and by phosphates is increasing, including pollution of ground water aquifers.[5]

Table 7. Planned Allocation of Subsidies by Seine-Normandie
Basin Agency, 1977-1981

FOR WATER QUALITY MANAGEMENT: 2300 x 10^6 francs[a]

	% of Total
Municipal treatment plants	20.5
Industrial treatment plants	19.8
Major interceptor sewers	11.5
Sewer networks	9.8
Other projects	13.4
Total for classical pollution	75.0
Facilities for reduction of discharges of toxic materials	13.5
Capital and operating costs of centers for treating solid toxic residuals	11.5
Total for toxic pollution	25.0
	100.0%

FOR WATER QUANTITY MANAGEMENT: 520 x 10^6 francs[b]

	% of Total
Aquifer protection	22.2
Dams and reservoirs	17.6
Paris region interconnections	27.8
Major water transfers	26.5
Other projects	5.9
	100.0%

[a]About 580 million dollars at 1978 exchange rate

[b]About 130 million dollars at 1978 exchange rate

- Thermal pollution will inevitably increase significantly because of the rapid development of the nuclear power program.[5]

- Bacterial pollution is developing in some places.

- Pollution by urban storm runoff has been given little consideration.

- In-plant modification and recycling is normally the first option to consider for reducing discharges, but inducements in that direction have not always been systematic nor possible.

- The interface with solid residuals management has often proved difficult to achieve.

- Discharges from agricultural and silvicultural operations are not even considered.

- Treatment of discharges from small human settlements, even though the technology exists, has often not been given adequate consideration.

- Even though the average efficiency of sewage treatment plant operations has increased from about 50% in 1970 to about 75% in 1978, further increase is needed in order to achieve the discharge objectives.

- The sewage and water distribution networks are, on the whole, totally inadequate. They now represent the limiting factor for the next step in decreasing discharges to water bodies. Because vast investments are necessary for these networks, e.g., about five times more than for treatment

plants, this represents a potential direction for expanding
the activities of basin agencies.

Given the indicated problems and trends, what is the shape of
the programs of the basin agencies likely to be in the last two
decades of the 20th century? First, the agencies will continue and
extend their classical operations with respect to water quality and
water quantity management, as enumerated above. However, this will
include taking into account new pollutants and new sources of those
pollutants. Second, the actions of most of the basin agencies will
be more and more oriented by the water quality objectives policy
which they already actively support and help to develop. This means
that better knowledge of the interactions between discharges and
ambient water quality in virtually all rivers and other water bodies
is necessary. It also means that a systematic procedure for deciding
on the standards for each discharger must be developed. Third,
because the water distribution and sewage collection networks represent
the limiting factor in achieving additional reductions in discharges
from point sources, the basin agencies may well increase substantially
their subsidies for such projects.

Finally, over the next two decades the assessment basis for
effluent charges will become much smaller than it was in 1978.[6]
(The assessment basis is comprised of the actual discharges--by
activities--of the four pollutants currently charged, i.e., TSS, OM,
soluble salts, toxics.) This will be true even though in the future:
(a) more residuals will be subject to charges; and (b) most of the
discharges from point sources are now actually measured and are no

longer assessed on a flat rate basis, which usually meant an under-assessment. Therefore, the level of charge per unit of discharge will continue to increase in real terms. Because fewer new investments will be needed, an high proportion of the revenues from the charges will be spent as premiums and superpremiums for those who treat their discharges. These premiums and superpremiums will represent a substantial help for operating costs of individual activities.

NOTES

[1] The program of expenditures is the multi-year "intervention" program developed by each basin agency. But this program must be checked and approved by the Basin Committee, the Delegated Basin Mission, and the Interministerial Mission on Water. Specific grants are checked by the executive board of each agency and the Basin Committee. The annual budgets of the agencies are approved by the National Assembly.

[2] Although this group is homogenous, this does not mean there are no deep conflicts within it. In fact, conflicts are all the deeper because the individuals belong to the same family. However, this homogeneity also means that a common position can quickly be achieved in face of a common threat.

[3] These zones are the same as the charge coefficient zones for water abstractions or wastewater discharges. They correspond to areas of different order of priority for action. For example, upstream areas are usually higher priority than downstream areas. See chapter 8.

[4] The objective is to push dischargers into seeking lower cost measures to reduce discharges.

[5] Effluent charges on nitrates, phosphates, and thermal discharges will be levied in the very near future.

[6] The assessment basis increased from the time of initiation of the charges in 1970 until 1978. Since then it has remained about constant.

Chapter 6

THE BASIN AGENCIES AND THE
WATER QUALITY OBJECTIVES POLICY

IMPLEMENTING THE WATER QUALITY
OBJECTIVES POLICY

The basis of the French water quality objectives policy is the 1964
Law on Water. The goal is to achieve, for all water courses, water quality
objectives within a given time horizon, which objectives are to provide the
basis on which decisions concerning investments and discharge authorizations
can be made. These objectives comprise the necessary basis for the authorization procedure. The water quality objectives are stated in terms of
"types of uses possible," which then imply specific levels for the various
water quality indicators.

On 29 July 1971, a decree was issued which specified the procedure
for implementing the water quality objectives policy. Further amplifica-
tion was provided by the decree of 17 March 1978, which stated two basic
procedures for implementing the policy.

The first procedure is aimed at establishing departmental maps of
ambient water quality objectives. The initial state of water quality must
be assessed, and the probable costs to the different economic activities to
achieve the objectives must be indicated. These maps must "give an image
of the desirable situation ten years from now, taking into account investment possibilities."

The responsibility for establishing the maps belongs to the departmental direction of agriculture or environment. The maps are drafted within the departmental directions, for each subbasin of the department. In that process, associations and users are informally consulted. At the end of this drafting stage, the Delegated Basin Mission--representing the basin in which the department is located--and the Technical Committee on Water of the region to which the department belongs, are consulted.

Next the formal consultation phase begins. This involves, among other steps, consultation with the general council of the department, the chamber of agriculture, the chamber of commerce and industry, the regional council and the Basin Committee. The outcome is the recommendation that the final map should be published and publicized. The procedure is supposed to be fairly rapid and "light."

The second procedure is much more detailed and complex. For example, several alternative maps of ambient water quality objectives must be submitted. The consequences of each of the objectives for the major dischargers must be calculated; public particpation is more extensive. The result is a map, but also a decree specifying the ambient water quality objectives. This procedure, the one which had been used prior to 1978, is formal, costly, and produced inflexible results because publication of a decree represents a very binding situation (including financial implications) for the national government.[1] Therefore this procedure simply could not be applied everywhere, so that a lighter, more flexible procedure has been adopted.

However, whichever the procedure, its basic stages are always the same.   They are:

(1)   identify the actual and desired alternative uses of water on each reach of the water bodies, with public participation;

(2)   given those uses, establish the corresponding ambient water quality required, i.e., through water use-water quality standards;

(3)   compute the maximum discharge compatible with the above established ambient water quality required, with the help of mathematical modelling such as use of the Streeter-Phelps formulation;

(4)   identify the critical water pollution parameters and the specific discharges which are responsible for the possible discrepancies between actual and desired ambient water quality;

(5)   given assumptions on the future development of the area, define the needed reduction in discharges and compute the corresponding capital and operating costs for the various activities involved (analysis and programming phase);

(6)   submit those "desired water quality costs" alternatives to the various actors and organize public debate about them;

(7)   promulgate the offical texts, e.g., water quality objectives decrees or prefectoral decrees specifying the

maximum admissible discharges, rewrite discharge

permits, etc.; and

(8) implement the policy at the administrative (authoriza-

tion procedure and control) and financial (grants)

levels.

By the end of 1978, several pilot operations had been launched

on subbasins in Normandy, Lorraine, and northeast of Paris.[2] These

operations have particularly involved the basin agencies. This is

because they play two key roles. First, they will subsidize many

of the activities which will be necessary to achieve the water quality

objectives. Second, they make most of the studies needed to prepare

the plans because they have the personnel, the data, and the knowledge

of the various technical possibilities, and have established

relationships with the major dischargers.

Departmental maps are now being produced. It appears that the

difference in level of detail of the maps and in extensiveness of

participation depends more on local circumstances and inclinations

than on the type of procedure used. The water quality objectives

policy is becoming the basis for action decisions at the local or

subbasin level, that is, specifically what degrees of discharge

reduction are to be achieved where. In effect, these decisions

give specificity to the multi-year aggregate programs at the basin level.

The water quality objectives policy, which is a subbasin level

policy under departmental control, has several major advantages. These

are the following.

- It increases the efficiency of investments by making the best use of the assimilative capacity of the rivers.

- It permits the elaboration of a coherent set of actions, specified in space and time, which define various stages towards the fulfillment of the ambient water quality objectives.

- It is the mechanism by which: (a) the concerned community balances and sets the tradeoff between the costs and benefits of improving ambient water quality; and

   (b) commits itself to reach the corresponding objectives.

- It allows for better and more realistic consideration of all factors influencing ambient water quality, such as nonpoint sources, landfills, accidental discharges, water temperature, low flow conditions.

- It leads to an analysis of the relationships between ambient water quality and land uses.

- It enables due consideration of complementary actions, such as scenic enhancement of river sites, protection of sand extraction areas, river bed cleaning, sludge disposal.

- It is the occasion of an intensive water quality measurement campaign, with respect to both ambient water quality and discharges, as well as an assessment of who is responsible for what pollution.

THE WATER QUALITY OBJECTIVES POLICY IN
RELATION TO REGULATORY AND ECONOMIC POLICIES

The water quality objectives policy contains potential contra-
dictions with the national discharge standards orientation of the
Classified Establishments Service. A change in attitude is
required by the SCE if water management decisions are to depend, at
least partly, on local conditions, as is implied by the water quality
objectives policy. In contrast to the national orientation of the
discharge standards of the Classified Establishments Service, the
activity-by-activity authorization procedure finds its justification
and rationale for existence in, and thus becomes a key component of,
the water quality objectives policy.

For the basin agencies, the water quality objectives policy
meant, if not contradiction, at least evolution. This policy implies
implementation of specific projects at specific places by certain
dates, despite the fact that the agencies cannot require anyone to
do anything, nor can they build anything themselves. Therefore,
in order to induce the desired actions, an agency would have to tailor
the incentives it has, subsidies and charges, to promote investments
in the most crucial areas in its basin. This would require differentiation of the economic incentives on virtually a case-by-case basis,
e.g., among activities and among small reaches of the same river.
However, thus far the differentiation has been, at best, on a gross
scale, such as division into upstream and downstream portions of a
basin. Achievement of the economically efficient degree of

differentiation is not likely because it is politically difficult to have significanly different subsidies and charges among communes in the same basin. However, the official policy is to increase effluent charges and the amount of subsidy in percent of capital costs in areas where ambient water quality standards have not been achieved.[3]

Two practical results have stemmed from the above factors. First, regardless of the zone, the level of the premium for treatment has been brought to its maximum level wherever the ambient water quality objectives have not been achieved, and to its minimum level where they have been achieved. This also means that a new criterion has been added for choosing projects to be subsidized. Second, regional "antennas" of the basin agencies have been established to provide closer contacts at the departmental and local levels. The regional antennas are subgroups of a basin agency, based permanently in major cities within the basin, e.g., subregional offices. Each is "in charge" of a set of subbasins and a set of departments more or less corresponding to those subbasins. The role of the regional antennas is to make and maintain contacts with the departmental administrations, industrial activities, and communes, and to prepare the ambient water quality objective plans.

The water quality objectives policy has the characteristics which probably make it an irreplaceable element of a general water management strategy. First, it brings the water users together to discuss and negotiate, which physically expresses the necessary solidarity of the various factions in a river basin. This is perceived to be the basis of sound water management in France. Second, it brings

legitimacy to subsequent financial efforts and sanctions. This is
the basis for real enforcement of legislation. Consequently, this
policy was needed in order to make further progress in water quality
management in France. However, to make such progress possible requires
adaptation and evolution in the regulatory and economic policies, and
a new type of cooperation and complementarity between them.

THE NEW COMPLEMENTARITY BETWEEN THE
REGULATORY AND THE ECONOMIC POLICIES

Although there traditionally has been close cooperation between
the basin agencies and the ministries at the national level, it is
fair to state that this has not always been the case at the departmental
or local levels. The policies were not necessarily conflicting, they
were just parallel, information flows being rather rare between these
two lines. This is quite understandable considering that certain
agencies have no financial but only regulatory and operating powers,
whereas the basin agencies have no regulatory or operating powers but
only financial powers. Further, some agencies, e.g., the Classified
Establishments Service, act on a national basis for each sector
of activity; others, e.g., the departmental directions, act on a
local-political basis; and the basin agencies act on a multi-regional
scale with a five-year time horizon.

Enforcement of legislation often lacked legitimacy, could be
suspected of being economically inefficient at the subbasin level,
and could easily look unfair. As a consequence, results of the regulatory policy have been quite mixed. On the other hand, the action of

the basin agencies could lack accuracy, be unable to tackle the real local problems, and lack "teeth" when needed. This situation is now evolving positively, because the agencies have now become mature and are widely recognized for their achievements, but also because the water quality objectives policy strongly requires it.

Thus, the departmental administrations and the agencies need each other more than before for two basic reasons. First, they must act together to establish the maps of water quality objectives, by law and by technical necessity. Second, the directing and inducing of actions and projects needed to reach those quality objectives need to be so accurate and coherent that they simply cannot do without each other's tools.

These necessities can be the basis for a new complementarity which has already shown itself in the recent initiation of the regional antennas of the agencies and the subsequent important role they have taken. Why shouldn't the national administration now have a coordinating mechanism at the basin level for the regulatory side of water quality management? If the future of the agencies lies in a better recognition of the specifics at the local level and in an extension of their roles of financier in relation to municipalities and departments, the future of the regulatory side of water quality management lies in the recognition of the necessities of legitimacy, equity, efficiency, and basin-level solidarity. The future of both lies in the deepening of cooperation, which is the key for success of the water quality objectives policy.

NOTES

[1]The problem is to know to what extent the national government
will consider itself bound by the water quality objectives which will
have been chosen by this procedure.

[2]In two years each department must have its map. No one knows
whether the status of a map will be: (a) a mere study; (b) a map
agreed upon by all parties; (c) a real plan; or (d) a prefectoral
decree. Given present trends, the map will probably be just "a study."

[3]See section on Standards and Criteria in chapter 4 and paragraph
on KZONE in chapter 8.

Section II

THE EFFLUENT CHARGES SYSTEM IN FRANCE

Section II is focused on the charges levied by the basin agencies.
The effluent charges are discussed in detail; only minimum attention
is given to charges on water intakes.

It should be emphasized at the start that the basic reason for
the charges is to raise money to achieve the multi-year programs of
the basin agencies. These programs consist of helping municipalities
and industrial activities finance their investments, by grants and
subsidized loans, and by subsidies for operating costs, to reduce their
discharges. In fact, neither the legal texts on which the charges
are based, nor the unit levels and structures of the charges, have
much to do with the concept of economic incentives.

Nevertheless, it is also true that the effluent charges have
multiple and often indirect effects on the decisions of the decentralized
actors regarding water use. These effects can be very different
among industrial sectors, among regions, and from one local sociopoli-
tical context to another.

The principal objective in this second section is to present the
effluent charge system in such a way that, given the general context
of the water management system presented in Section I, hypotheses can
be made about, and an assessment can be made of, the incentive power
of the charges, and, more generally, the role of effluent charges
in water quality management in France.

Chapter 7

<u>THE LEGAL BASIS FOR THE EFFLUENT CHARGE SYSTEM,</u>
<u>MONETARY FLOWS IN RELATION TO EFFLUENTS, AND</u>
<u>A FIRST ASSESSMENT OF THE SYSTEM</u>

THE LEGAL BASIS FOR THE
EFFLUENT CHARGE SYSTEM

Both laws and interpretation of laws have served to establish
the legitimacy of the effluent charges system in France.

<u>The Legal Texts</u>

Article 14 of the 16 December 1964 law states: "The Basin Agency
sets and receives payment of charges from public or private bodies
when these public or private bodies make the intervention of the Agency
necessary or useful, or when they have profited by the interventions."
Article 18 of the 14 September 1966 decree, promulgated in application
of the 1964 law and modified later by the decree of 28 October 1975,
specifies that: "Charges can be demanded from public or private bodies
which make the intervention of the (Basin) Agency necessary or useful
because they (the bodies) contribute to deterioration of water quality,
because they withdraw water, because they modify water flows. Fees can
also be demanded from public or private bodies which benefit from works
or investments made with the aid of the Agency."

The Legal Status of the Charges[1]

In order to establish practically the charges system, the minister in charge of economic and land use planning--which at that time (1966) had the tutelage of the basin agencies--asked the Council of State to specify the legal status of the charges. The answer, given on 27 July 1967, was that given various legal considerations, the charges were neither ordinary taxes, nor special taxes (taxes parafiscales), nor payment for a service. Therefore, the Council of State concluded that the 1964 law had created a new type of charge "corresponding to the specific nature of the operation."[2]

The first major consequences of the Council's response are the following. One, there is no link whatsoever between the charge paid by a discharger and the aid it can receive; payment of the charge does not entitle the discharger to any aid and receipt of aid is not contingent upon paying a charge.[3] Two, the charges levied by the basin agencies do not have to take into account the various exceptions attached to existing charges and taxes for all types of categories and activities. Three, nevertheless, the charges are public and have official status. This means that all the measures used to collect other taxes and charges imposed by public agencies can be used, when necessary, to collect the charges levied by the basin agencies.

The Interpretation of the Texts

The legal texts establishing the charges system leave several questions open. Examples include: can an activity be charged both for discharging and for benefiting from improved water quality as a result

of upstream wastewater treatment made with the financial aid of
the basin agency? What are the relationship(s) between charges for
discharge of pollutants (residuals) and charges for water intake,
because these two types of water use interact with each other?

In practice, the texts have been interpreted so that the charges
system would be as simple as possible. An effluent charge has been
set and a water intake charge has been set, with both being levied on
an activity without considering whether the activity receives any direct
benefit from any action undertaken by a basin agency. Further, only an
indirect link has been kept between the level of the charge and the
expenses incurred by a basin agency, by establishing different zones within
a basin. Among these zones the following differ: (a) the unit levels for
the effluent charge on a specific pollutant; (b) the maximum costs per
unit of wastewater treatment admissible to receive financial aid; and
(c) the proportions of aid given by the basin agency.

The effluent charges system has been effectively working since
1969. The final set of decrees organizing all its practical aspects
was issued on 28 October 1975. Between 1969 and 1975, the system ran
on a temporary basis for charges on municipal discharges, because of
a legal challenge on the part of the Association of the Mayors in
France, which was opposed to paying the charges. The case was settled
by changing the procedure for collecting the effluent charges to which
the municipalities are subject, so that subsequently the effluent
charges system has been operating regularly and reasonably efficiently.

MONETARY FLOWS IN
RELATION TO EFFLUENTS

Activities, Other than Municipalities, Discharging Directly into Water Bodies

For individual activities which do not withdraw from or discharge to municipal water distribution and sewage systems, there are two situations, as depicted in figure 11: one where the activity does not have a treatment plant; the other where the activity does have a treatment plant. In the first situation, either of two procedures is used. One, the individual activity, e.g., an industrial plant, pays to the basin agency an annual effluent charge (redevance de pollution) on its generation of pollutants,[4] based on the assumption that generation equals discharge. The generation/discharge coefficient is taken from a table, as described in chapter 8 and illustrated in table 10. (The activity also pays an annual water intake charge [redevance de prelevement d'eau].) Alternatively, either the activity or the agency requests that an actual measurement of the discharge be made, in which case the annual effluent charge is based on the measured discharge. Because there is no reduction in discharge by a treatment plant, the activity receives no superpremium.

In situations where an activity has installed and operates a treatment plant, the quality of the effluent is measured and the annual effluent charge is based on the measured discharge. Depending on the efficiency of the wastewater treatment plant, e.g., degree of reduction in discharge of pollutant(s) actually achieved, as determined by measured influent and measured effluent, a superpremium (sur-prime or incitation

Figure 11.   Bases for Calculating Effluent Charges for
             Activities Discharging Directly to Water Bodies

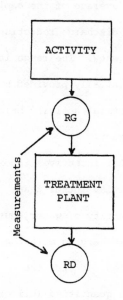

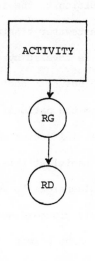

Activity pays effluent
charge on actual dis-
charge; receives super-
premium depending on
efficiency of treatment
plant.

Where activity has no
treatment plant, generation
is assessed equal to discharge.
Activity pays effluent charge
based: (1) on measurement of
RD; or (2) on coefficient
taken from table. No superpremium is paid.

Abbreviations: RG = residuals generation

                RD = residuals discharge

a la depollution maximale) is paid by the basin agency to the activity
at the end of the year.[5] The activity must have an approved system
for disposal of sludge before the superpremium is paid.

In addition to the superpremium, 30% on the average of the capital
costs of wastewater treatment or other pollutant discharge reduction
facilities is covered by grants from the basin agency. The range is
from 10% to 50%. In addition, another 10% is eventually provided by
a grant from the national government if the industrial activity belongs
to an industrial sector which has a sector contract.

It is useful at this point to review the other incentives imposed
on direct discharges to water bodies. All activities (other than
agricultural) discharging more than a certain quantity of water, must
have a discharge permit. For those activities discharging more than
this minimum and belonging to one of the classified sectors, the
discharge permit specifies, at a minimum: (1) the quantities (kilograms)
of TSS, BOD_5, salts, and toxic materials which can be discharged per
ton of product produced, and the maximum quantities which can be discharged per day; and (2) that an appropriate discharge sampling location
must be constructed and provided. Self-monitoring and reporting of
discharges can be required.

Municipalities and Activities Linked to a Municipal
Water Distribution and/or Sewage System

For municipalities, and for all industrial, commercial, and service
activities with water intake of 6,000 cubic meters (about 1.5 million
gallons) or less per year, and which are linked to a municipal water
distribution and/or municipal sewage system, the monetary flows are

more complex. This is illustrated by describing the general case
where the activity is an household; the water distribution system is,
as is typical, owned and operated by a private water distribution
company (or private water company); and the sewage system (collection
and treatment) is owned and operated by a municipality or, as is
often the case, by several municipalities jointly.

The Charges

The water intake charge is paid to the basin agency by the
water company and is part of the water price which is billed by
the company to the individual household (and other activities) on
the basis of its quantity of water intake. The effluent charge is
not paid by the municipality to the basin agency, contrary to what
had been decided at the initiation of the system in 1969. Instead,
the municipality's effluent charge is paid by the individual household,
individual commercial activity, individual industrial plant (i.e., all
activities discharging to the municipal sewage plant) to the water
distribution company along with the individual household's water
bill, as an increase in the price per cubic meter of water intake.
This money, after three months, is then paid by the water company
to the basin agency.

At the beginning of each year, knowing the population of the municipality and the level of the effluent charge per inhabitant-equivalent
from the basin agency--one inhabitant-equivalent corresponding to 90
grams per day of suspended material and 57 grams per day of oxidizable
material--the total effluent charge for the municipality is computed.

Knowing approximately the volume of water intake in the municipality for the year, the additional price per cubic meter of water is then computed, to yield the amount of money corresponding to the effluent charge--the counter-value (contre valeur)--to be paid by the water users in the municipality. (Municipalities with populations less than 400 inhabitants and agricultural activities do not pay effluent charges.)

The operator of the sewage treatment plant may receive, directly from the basin agency, a premium based on the amount of pollutants removed in treatment. This amount is based either on the assumed quality of influent (load) to the plant, i.e., 90 grams of suspended material and 57 grams of oxidizable material per inhabitant per day, and the measured amounts of these materials in the effluent, or on measurements of both influent to, and effluent from the treatment plant.[6] The operator of the treatment plant may also receive a superpremium from the basin agency based on the performance of the treatment plant. There must be an approved system for the disposal of sludge before the premium and superpremium are paid to the operator of the treatment plant.

It should be noted that, because the premium and superpremium together usually cover less than half the operating expenses, the incentive still is not to use it, even if the treatment facility is already in place.[7]

The Sewage Tax

To cover: (1) the part of the operating costs of the sewage system
not covered by the premium plus the superpremium; (2) the part of the
capital costs (debt service) of the sewage treatment facilities not
covered by grants from the basin agency and the department; and
(3) the capital costs of the sewage collection system which are not
covered by the general council,[8] the municipality levies a sewage
tax on each activity, e.g., households, commercial enterprises. (The
amount of this tax per cubic meter of water must be approved by the
prefect.) Here again, the sewage tax, which is much more important
than the effluent charge in terms of magnitude, is usually collected
by the water distribution company through an additional price per
cubic meter of water intake. The decree of 5 January 1970, which
specified this procedure, specified that fire stations, public fountains,
etc., and agricultural activities do not have to pay the sewage tax.

For activities with an annual water intake of more than 6,000
cubic meters, including hospitals and public buildings, there is a
declining tariff schedule. The sewage tax is divided by two for
activities with intakes of more than 50,000 cubic meters (about 13
million gallons) per year.[9] Furthermore, if the quality of the effluent
of such an activity is significanly different from that of an household,
a measurement of the effluent quality is made or a flat rate quality
coefficient is used which has been made official through a prefectoral
decree. The amount of the effluent charge for the activity is then the
charge which would have been paid by the activity if it had not been

linked to the municipal sewage system, i.e., if it had discharged
directly to a water body.

It should be emphasized that the decree of 10 June 1976 specifies
requirements for an effluent from an industrial activity to be accepted
in a municipal sewage system. These pretreatment standards are
analogous to the discharge standards imposed on activities with
permits to discharge directly to water bodies. Thus, whether the
discharge is to a municipal system or directly to a water body, both
effluent charges and discharge standards are imposed on the discharge.

As noted above, the sewage tax is a far more significant source
of revenues (funds) than the effluent charge. Thus, for an industrial
activity discharging into a municipal sewage system, its effluent
charge represents about 10% of its total payments to the sewage system,
i.e., the sum of the effluent charge and the sewage tax. For a municipality, revenues from effluent charges represent 10-15% of sewage
tax revenues.

Effluents and the Municipal Budget

The decree relating to effluent charges requires that the financial
situation of the sewage collection and sewage treatment facilities be
balanced. The operating costs are basically to be covered by the sewage
tax and the premium and superpremium. The capital costs are to be
covered by: (1) grants from the basin agency and the general council;
(2) subsidized loans from the bank called the Caisse des Depots
et Consignations;[10] and (3) the sewage tax. However, because there
is an upper limit on the level of the sewage tax per cubic meter of

water intake, it can well happen (and this is often the case) that a good portion of the debt service on the capital costs not covered by grants has to be covered by funds from the general budget of the municipality.[11]

It must also be stated that the accounting for depreciation and interest for public investments such as sewage collection and sewage treatment facilities, and their positions in the budget are often not quite clear. On the whole, there is an underevaluation of these costs. Finally, it is not always clear who pays for what in relation to these costs.

Financing of the management of storm water runoff is part of the general municipal budget. When some of the storm water runoff is handled in the sanitary sewer system, a contribution to the costs of the sanitary sewage system is made according to the degree of separation between storm runoff and so-called sanitary and process effluents. This contribution is between 10% and 50% of capital and operating costs of the requisite facilities.

In table 8 are summarized the costs to be covered by the municipality and the sources of funds to cover those costs. All of the local decisions involved are systematically made under the control of, and with the required approval of the prefect, who often takes advice from the various concerned department directions.

These questions of the financial aspects concerning effluents from the municipal systems will be more and more important in the future because the bottleneck for reducing discharges into the environment from municipalities, and activities in municipalities, is now more the

Table 8.   Annual Costs of Municipal Sewage, and Associated
            Payments by, and Sources of Funds to, Municipalities[a]

Costs to be Covered by Municipality:
Debt Service[b] and Operation and Maintenance

Sewage collection system
Sewage treatment system
Sewage treatment sludge handling and
    disposal system

Annual Sources of Funds to Municipality	Annual Payments by Municipality
Sewage taxes	Debt service on all facilities
Premiums and superpremiums from basin agency	Operation and maintenance costs of all facilities
General municipal revenues	

[a]
All costs allocated to handling storm water runoff are paid from
general municipal revenues.

[b]
Total capital costs minus grants from the basin agency and the
department represent the magnitude of the debt to be serviced
by the municipality.

insufficiency of sewage collection systems than of sewage treatment

plants. Because a municipal sewage collection system in France costs

from 5 to 7 times more than a sewage treatment plant, decisions must

now be taken to the national level with respect to the future directions

of actions of the basin agencies. These decisions include: what could

be the role of the basin agencies in the financing of sewage collection

systems? How will capital costs of facilities really be accounted for?

What interaction between municipal general budgets and sewage treatment

budgets is considered sound? To what extent should the general
municipal budget finance sewage collection and sewage treatment?

These questions represent some of the major questions about public
investments and the role of municipalities. Since 1969 when the effluent
charge system was initiated, the institutional context of water management in France has changed. The effluent charges system has brought a
new partner, the basin agency, into the game, a game which had been
stagnant for a long time. The new system might provide the impetus
to move ahead. Thus, an assessment of the system after its first ten
years is appropriate.

A FIRST ASSESSMENT OF THE
FRENCH EFFLUENT CHARGES SYSTEM

The French effluent charges system has two basic characteristics.
One, the charges are levied by the basin agencies and the revenues
are redistributed to dischargers by means of: (a) grants and subsidized
loans, together amounting to an average of 30% of the capital costs of
facilities to reduce discharges; and (b) the premiums and superpremiums
for operating costs. Two, the charge level per unit is 1/2 to 1/3
of the average annual operating costs of a treatment plant, per unit
treated. Starting with these characteristics, this section contains
a brief analysis of where the effluent charges system stands regarding:
(1) what is really paid for by the dischargers; (2) the efficiency of
the system; and (3) the relationship of the system to the polluter-pays
principle.

For What Do the
Dischargers Really Pay?

In the French case, no payment is made for compensation for residual damages nor for the use of a common property resource. In paying effluent charges, dischargers are providing funds to be used for financial aid granted to dischargers which supposedly have the lowest unit treatment costs in a given subbasin. The levels of treatment are defined in relation to a general discharge reduction objective, as expressed in the basin agency's multi-year program, and in relation to more specific ambient water quality objectives, as reflected in the departmental water quality objective maps.

The Efficiency Question

Given the low level of the effluent charges, nothing in economic theory guarantees the efficiency of the resulting resource allocation. In fact, there are three reasons why the existing system appears to achieve some degree of efficiency. First, even with the relatively low levels of the charges, activities where it is cost-effective to undertake some discharge reduction find it profitable to do so.[12] Second, those industrial sectors where it is more cost-effective to reduce discharges have been pushed, at the national level, to do so with the sector contracts. Third, the basin agency, both through knowledge of the individual dischargers and through its criteria for grants,[13] makes sure that the resource allocation is as efficient as possible. The resource allocation is not as good as in the case of a truly incentive charge,[14] because the basin agency cannot know exactly the marginal cost

of reduction of each discharger. But it is not as bad as it would be
in the case of a general standard, such as 80% reduction by all dischargers.

Relationship to the
Polluter-Pays Principle[15]

Two of the major points mentioned in the recommendations of the
Council of the Organization for Economic Cooperation and Development
are:

- The polluter-pays principle simply means that
  the polluter is responsible for all the pollution
  control operations required to achieve the objectives
  set by the authorities, whatever they may be.

- To give aid to those who treat wastes conflicts
  with the polluter-pays principle "unless the
  transfer payments correspond to the purchase
  of a service from users who treat wastes more
  thoroughly than usual due to their greater
  efficiency."

Consistency with the polluter-pays principle therefore means that:
(1) there must exist an ambient water quality target, standard, or
objective of some kind; and (2) to reach that target, dischargers directly
pay the costs to meet it in the case of individual standards, or make
a payment (effluent charge) which corresponds to the purchase of
sufficient services from dischargers which have lower discharge
reduction costs, so that the overall ambient water quality standards
are met.

In the early 1970s, the effluent charges were very ineffective and
the discharge standards on the individual dischargers were either
disregarded or were out-of-date, i.e., old prefectoral authorizations
reflecting out-of-date political choices. Under these circumstances
it could be argued that: (1) the objective was unclear and anyway, not

given due consideration; and (2) such low charges could hardly correspond to the purchase of any service. The charges looked more like the purchase of a right to pollute and to disregard the law, which has little to do with the polluter-pays principle. Now, at the end of 1979, the levels of the effluent charges in the various basins have significantly increased and the ambient water quality objectives are clearly specified reach by reach, as part of the regional/national ambient water quality objectives policy. However, there is often an unclear situation regarding the degree of compliance with individual discharge standards, which themselves are not necessarily consistent with the ambient water quality objectives policy. But the actual structure and levels of effluent charges, and the system of grants and aids, appear to be reasonably consistent with the polluter-pays principle.

NOTES

[1] For a detailed analysis of this question see Nicolazo-Crach, J.-L. and Lefrou, C., 1977, Les Agences Financieres de Bassin, P. Johanet, Paris, pp. 130-134.

[2] Semantics remain a problem. Charge, fee, tax are three terms variously used. However, in the U. S. context, "tax" has a particular legal meaning in relation to raising of revenues by governmental bodies. A charge is normally considered to be imposed as payment for a service rendered. A fee, as a license fee or abstraction fee, is often a payment to obtain permission to undertake some action. It does not have any necessary relation to the value of the material or service involved. The term "charges" has been used throughout this report.

[3] One could imagine a discharger who does not pay a charge because his discharge is under the minimum level, but who would be aided by a basin agency to reduce his discharge.

[4]This payment is due on 1 January for the current year.

[5]December 31 of the current year.

[6]Depending on the mix of activities in the municipality, the difference between the assumed value of an inhabitant-equivalent and the actual value, in terms of influent to the treatment plant, may be quite large.

[7]See chapter 8.

[8]For this portion of the costs, the revenue from the sewage tax is not always sufficient. In such cases funds from the general budget can be used as well as special borrowing.

[9]This procedure shows that the sewage tax is mainly concerned with quantity of water and not quality of water; only the effluent charge is concerned with quality. The dominance of the sewage tax means that some significant portion of the costs resulting from discharges of TSS and OM is shifted to quantity discharges.

[10]The bank is a public institution which obtains money from the national savings account system operated by the post office, and lends it to municipalities and general councils at a rate well below the market rate.

[11]This is one of the reasons why some municipalities refused to pay what appeared to them to be a new tax when the effluent charge system was established.

[12]The level of the charge sometimes is sufficiently high for it to be an incentive for some dischargers to reduce discharges.

[13]For aid to be granted, there is a lower limit on the expected discharge reduction per franc spent. This limit varies among the zones in a basin, which zones represent different levels of needed discharge reduction.

[14]See chapter 8.

[15]For a more complete discussion of this principle, see Barde, J. P., 1975, "An examination of the polluter-pays principle based on case studies," in The Polluter-Pays Principle, OECD, Paris, pp. 93-117. The two quotations in the first paragraph of this section are from page 94.

Chapter 8

STRUCTURE AND COMPUTATION OF
WATER INTAKE CHARGES AND EFFLUENT CHARGES

INTRODUCTION

Each year since 1969, each activity--whether discharging directly
to a water body or into a municipal sewage system--which uses more
than 6,000 cubic meters (m^3) of water per year has received a water
intake bill and an effluent charge bill. Each household or other
activity linked to a municipal sewage system and using 6,000 cubic
meters of water or less per year, has seen two new items on its water
bill: a water intake charge and an effluent charge.[1] The water
intake charge is based on the volume of the water intake, and applies
to surface waters as well as to ground waters. For activities using
more than 6,000 m^3 of water per year, the effluent charge is based
on the contents in the discharge of suspended material (TSS),
oxidizable material (OM), salts (in essence, total dissolved solids,
TDS), and toxics. For activities using 6,000 m^3 or less of water
per year, the effluent charge item on the water bill simply reflects
the total effluent charge for the municipality divided by the total
water intake of the water users in the municipality (the unit charge),
multiplied by the water intake of the activity.

COMPUTATION OF WATER INTAKE CHARGES

The water intake charge is based on a unit charge from a table
which shows the charge per cubic meter of water abstracted. This
unit charge is a function of four factors. The first factor is
whether the abstraction is from surface or ground water. Unit
charges are generally higher for ground water than for surface water
and even higher for specific aquifers, such as the Albien aquifer
in the Paris region.

The second factor is the zone in which the abstraction takes
place. These zones reflect special protection areas for aquifer
recharge and areas which have required specific investments by the
basin agency. For example, there are nineteen zones within the area
of the Seine-Normandie Basin Agency. These water abstraction-related
zones are different from the effluent charge zones.

The third factor is the time of year. Unit charges are usually
higher during low-flow periods.

The fourth factor is whether the water is returned to the
environment after use by the activity, or is consumed and not returned
to the environment in the basin. The proportion of water consumed
for a given amount of water intake defines the restitution coefficient,
which depends on the type of use of water. The unit water consumption
charge is based on this quantity. The water intake charge (redevance
ressource) is the sum of the water abstraction charge (redevance de
prelevement) and the water consumption charge (redevance de consom-
mation).

The above factors are reflected in the unit charges shown in table 9. The rationale behind this schedule of charges is the management of scarce water supplies. Thus, charges on surface water abstractions are levied only in low flow periods; charges, generally substantially higher, are levied year around on aquifers to which the water is not returned. Similarly, the unit charges on consumptive use of water are generally several times higher than the unit abstraction charges.

COMPUTATION OF EFFLUENT CHARGES

For Activities Other Than Municipalities

For activities which are not linked to a municipal sewage system and have no treatment plant, i.e., discharge directly to water bodies, and ostensibly[2] for activities which are linked, but have an intake of more than 6,000 cubic meters of water per year, the annual effluent charge is computed according to the following formula:

$$TECHARGE\ (P) = VOL \times KDIS\ (P) \times KZONE \times BASE\ (P)$$

where, TECHARGE = total _annual_ effluent charge for pollutant P;

VOL = volume of production in physical units (grandeur caracteristique), implicitly per day;

KDIS (P) = specific coefficient (coefficient specific), i.e., the amount of pollutant P discharged per unit of volume of production as actually measured, or assigned from a table, assuming generation = discharge;

Table 9.   Unit Charges on Water Abstractions and Consumptive
Use of Water, by Zone in the Seine-Normandie Basin, 1971[a]

Zones	Surface Water			Ground Water		
	Operation	From 1st June to 31st Oct.	Rest of Year	Operation	From 1st June to 31st Oct.	Rest of Year
1.1	Abstraction[b]	0.200	0	Abstraction other than Albien	7.0	7.0
	Abstraction[c]	7.000	1.0	Abstraction Albien	11.0	11.0
1.2	Abstraction[d]	0.030	0	Abstraction other than Albien	7.0	7.0
				Abstraction Albien	11.0	11.0
2.0	Abstraction	0	0	Abstraction	1.0	1.0
	Consumption	3.500	0	Consumption	3.5	0
2.1	Abstraction	0	0	Abstraction	1.0	1.0
	Consumption	3.000	0	Consumption	3.0	0
2.2	Abstraction	4.000	0	Abstraction	4.0	0
2.3	Abstraction	0.100	0	Abstraction	4.0	4.0
	Consumption	3.500	0	Consumption	3.5	0
3.1	Abstraction	0.100	0	Abstraction	5.5	5.5
3.2	Abstraction	0.100	0	Abstraction	10.0	10.0
3.3	Abstraction[e]	6.500	6.5	Abstraction	6.5	6.5
3.4	Abstraction	0.100	0	Abstraction	9.0	9.0
4.1	Abstraction	0.025	0	Abstraction	1.0	1.0
	Consumption	0	0	Consumption	3.0	3.0
4.2	Abstraction	0.025	0	Abstraction	3.0	3.0
4.3	Abstraction	0.025	0	Abstraction	1.0	1.0
	Consumption	0	0	Consumption	11.0	11.0
4.4	Abstraction	0.025	0	Abstraction	11.0	11.0
5.0	Abstraction	0.025	0	Abstraction	1.0	1.0
5.1	Abstraction	0.025	0	Abstraction	3.0	3.0
5.2	Abstraction	0.025	0	Abstraction	1.0	1.0
				Consumption	2.0	2.0
5.3	Abstraction	0.025	0	Abstraction	2.0	2.0
5.5	Abstraction	0.300	0	Abstraction	2.0	2.0
	Consumption	3.500	0	Consumption	3.5	0

[a] All values in centimes per cubic meter

[b] Abstraction followed by discharge near the point of abstraction above the Seine-Oise junction, using special pipes and not the public sewers

[c] Other abstraction above the Seine-Oise junction

[d] All abstraction below the Seine-Oise junction

[e] Excluding abstraction from the Seine and the Tancarville Canal

KZONE = zone coefficient (coefficient de zone), the magnitude of which depends on the location of the activity within the basin;

BASE (P) = base charge coefficient (coefficient de base) for `pollutant P, for the given basin, francs per unit of pollutant P; and

VOL x KDIS (P) = amount of pollutant P discharged daily by the activity, also called the assessment basis (assiete de redevance).

The total annual effluent charge to be paid by the activity is the sum, for all relevant pollutants, of the charges for each of the pollutants. The variables in the formula are discussed in the following paragraphs.

VOL. The assessment basis for this variable is defined in article 3 of the decree of 28 October 1975 as being the "normal day of the month of maximum discharge activity." The volume of production is therefore calculated by dividing the total production in the month of maximum production in the year by the number of days of production in that month.

KDIS (P). There are two elements in the development of discharge coefficients. The first element involves the definitions of the pollutants for which charges are assessed, e.g., units of measurement. With respect to the specific pollutants, for suspended solids (TSS), the unit is the weight of suspended material. For oxidizable material (OM), the unit is the weight of oxygen required for the decomposition of the matter over a specified time period. The unit is separated into chemical oxygen demand (COD) and five-day biochemical oxygen demand (BOD_5), to which have been assigned the weighting coefficients of 1/3 and 2/3, respectively. The weight of OM is therefore computed as:

$$OM = \frac{COD + 2\ (BOD_5)}{3}$$

For salts, the unit is expressed in terms of conductivity in micromhos/m^3 multiplied by the cubic meters per day of soluble salts (TDS). For toxics, the unit is based on inhibition of life as measured by toxicity tests on daphnids. Toxicity is measured in kilos of equitox per day of inhibiting substances. If a raw untreated effluent kills 50% of the daphnids, it contains by definition 1 equitox per cubic meter; if it has to be diluted 50 times before the level of 50% kill of daphnids is reached, then the effluent is said to contain 50 equitox per cubic meter.

Having defined the units of pollutants, the second element involves the determination of the number of units of each relevant pollutant in the individual discharge. Two procedures are used. The first is to measure the discharge directly. According to the 28 October 1975 decree, the discharge coefficients, and hence the discharge measurement, must correspond to a "normal day" in the month of maximum activity. Usually when the basin agency makes a discharge measurement, that measurement is considered to represent a "normal day." However, it is not very clear what happens if one of the two parties--the discharger or the agency--contends that, in fact, the day of measurement was not a "normal day." It has been suggested that, in case of disagreement, measurements on three other days could be made and that the second worst day would be chosen as the "normal day."

The October 1975 decree also states that either the discharger or the basin agency, whenever either wants, can demand that a measurement of discharge be made. The party demanding the measurement must pay for the cost of the measurement. Since 1970, the basin agencies have budgeted an annual money appropriation for those measurements and have regularly and steadily demanded and made them. (The basin agencies have made many more demands for discharge measurements than have the industrial activities.) Hence, by the end of 1979, the majority of direct dischargers--in terms of volume of discharge, not necessarily in number of plants--have had their coefficients determined by actual discharge measurements. For example, by 1983, 95% of all discharges in the Seine-Normandie basin will have been assessed through measured coefficients. However, it should be noted that there is no annual or regular measurement. Rather, the discharge coefficients developed from a given measurement are valid until one of the parties asks for a new measurement.

The other procedure is to use a table in the 28 October 1975 decree, which shows, for 285 industrial activities: (1) the unit by which the volume of activity must be measured, e.g., physical units of production, volume of a specific input material, number of employees; and (2) the coefficients per unit of activity for the four pollutants, TSS, OM, salts, and toxics. A section of that table (tableau d'estimation forfaitaire) is shown as table 10. When the table is used, it is assumed that generation equals discharge.

KZONE. All basin agencies have divided their territories into various zones, for each of which the base charge is multiplied by a

Table 10. Examples of Coefficients Used in Establishing Effluent Charges for Individual Dischargers[a]

Category of Activity	Unit	Generation Coefficients		
		TSS[b]	OM[c]	Toxics
Production of pulp for paper				
<u>With destruction of black liquor</u>				
Unbleached kraft		10.0	40.0	0.21
Bleached kraft		40.0	90.0	0.35
Semi-chemical		40.0	90.0	-0-
Bisulfite	kg of paper product, 10% moisture	50.0	250.0	1.60
<u>Without destruction of black liquor</u>				
Unbleached kraft		20.0	240.0	d
Bleached kraft		50.0	290.0	d
Semi-chemical		50.0	290.0	d
Bisulfite		60.0	450.0	d
<u>Manufacture of artificial and synthetic fibers</u>				
Manufacture of viscose		28.0	35.0	2.50
Manufacture of other artificial fibers	kg of product material	9.5	7.5	2.50
Manufacture of synthetic fibers		9.5	7.5	-0-

[a]Discharge assumed equal to generation

[b]Total suspended solids

[c]Oxidizable material

[d]To be based on measurement

coefficient varying from 0.4 to 1.5, and averaging about 1.1. The designation of the zones and the assignment of coefficients to the zones depends on local conditions and on the policy of the basin agency. For example, aquifer recharge areas or upstream areas with relatively high ambient water quality have a coefficient of 1.5; heavily polluted industrial regions tend to have a coefficient of 1.0 or less. On the seashore, the coefficient depends on the nature of the dominant activity, e.g., industrial or tourism, and whether or not there is commercial fishing or shellfish culture.

BASE (P). The base charge for any given pollutant is determined through a technical, political, and administrative process described in the next chapter. The base charges for a given pollutant vary from basin agency to basin agency. For example, in 1978, the base charge for oxidizable material was 46 francs per kilogram in the Adour-Garonne Basin Agency, compared with 103.5 in the Seine-Normandie Agency. Per inhabitant-equivalent, the base charges were 5.58 and 10.0 for the same two agencies, respectively. During the period 1970-1978, the base charges have about doubled in constant francs, and quadrupled in current francs. Table 11 shows, for all six basin agencies, the evolution of the base charges from 1969 to 1983, for inhabitant-equivalents, TSS, OM, and toxics. Because the base charges are linked to the multi-year action programs, as described in the next chapter, the base charges are known for the next four years as soon as the votes on the action programs have taken place.

For activities discharging into municipal sewage systems, there must first be a determination of whether any given activity will be

Table 11. Evolution of Effluent Charges of Basin Agencies, Historical and Projected, 1969–1983[e]

Basin Agency[a]	1969	1970	1971	1972	1973	1974	1975	1976	1977	1978	1979	1980	1981	1982	1983
Francs/I.E.[b]															
A.G.	1.20	1.50	1.80	2.50	2.50	3.50	4.25	5.00	5.32	5.58	6.31	6.31	7.14	7.14	-
A.P.	2.00	2.00	2.00	2.00	2.00	2.60	2.60	3.92	4.18	6.12	7.34	8.57	8.57	8.57	8.57
L.B.	1.20	1.50	2.25	2.77	3.30	3.30	4.62	5.84	6.22	7.70	8.77	10.09	10.09	10.09	10.09
R.M.	3.00	3.90	3.90	3.90	3.80	5.10	5.10	7.65	7.65	7.70	7.65	7.65	7.65	7.65	7.65
R.M.C.	2.16	2.70	2.70	2.70	3.80	4.60	4.60	7.00	7.45	7.65	7.65	7.65	9.50	9.50	9.50
S.N.	2.20	2.20	2.20	3.75	4.41	4.41	4.41	7.75	8.25	10.00	11.55	11.55	11.55	11.55	11.55
Francs/kg TSS[c]															
A.G.	8.15	10.20	12.30	17.00	17.00	23.80	28.90	34.00	36.21	34.00	34.00	34.00	35.00	35.00	-
A.P.	11.00	11.00	11.00	11.00	11.00	14.30	14.30	22.00	23.43	30.00	36.00	42.00	42.00	42.00	42.00
L.B.	8.15	10.20	15.30	18.57	18.87	18.57	31.43	39.73	42.31	52.41	59.64	68.61	68.61	68.61	68.61
R.M.	16.70	21.70	21.70	21.70	21.70	28.40	28.40	42.60	42.60	37.50	37.50	37.50	37.50	37.50	37.50
R.M.C.	8.27	10.33	10.33	10.33	14.55	17.60	17.60	26.80	28.54	34.50	36.40	35.40	37.50	36.40	36.40
S.N.	15.00	15.00	15.00	25.50	30.00	30.00	30.00	33.00	40.47	49.00	56.62	56.62	56.62	56.62	56.62
Francs/kg OM[d]															
A.G.	8.15	10.20	12.30	17.00	17.00	23.80	28.90	34.00	36.21	46.00	57.00	57.00	70.00	70.00	-
A.P.	17.50	17.50	17.50	17.50	17.50	22.75	22.75	34.00	36.21	60.00	72.00	84.00	84.00	84.00	84.00
L.B.	8.15	10.20	15.30	18.87	18.87	18.87	31.43	39.73	42.31	52.41	59.64	68.61	68.61	68.61	68.61
R.M.	26.30	34.20	34.20	34.20	34.20	44.30	44.80	67.20	67.20	75.00	75.00	75.00	75.00	75.00	75.00
R.M.C.	24.80	31.00	31.00	31.00	43.65	52.80	52.80	80.40	85.62	103.50	109.20	109.20	109.20	109.20	109.20
S.N.	15.00	15.00	15.00	25.50	30.00	30.00	30.00	76.00	80.94	98.00	113.24	113.24	113.24	113.24	113.24
Francs/kg toxics															
A.G.	-	-	-	-	-	612.5	612.5	600	640	650	700	700	800	800	-
A.P.	-	-	-	-	-	.-	500	575	612	770	847	932	1025	1127	1240
L.B.	-	-	-	-	-	600	600	720	765	875	940	1080	1080	1080	1080
R.M.	-	-	-	-	-	.-	600	600	680	1000	1000	1200	1200	1200	1200
R.M.C.	-	-	-	-	-	400	400	560	596	720	760	760	760	760	760
S.N.	-	-	-	-	-	800	800	800	852	1000	1150	1150	1150	1150	1150
Price Index, Base 100 = 1970	95.0	100.0	105.5	112.0	120.2	136.7	152.8	167.0							

[a] A.G., Adour-Garonne; A.P., Artois-Picardie; L.B., Loire-Bretagne; R.M., Rhin-Meuse; R.M.C., Rhone-Mediterranee-Corse; S.N., Seine-Normandie

[b] Inhabitant-equivalent

[c] Total suspended solids

[d] Oxidizable material

[e] These charges are paid annually on the basis of a daily rate of discharge as explained in the text, that is, the discharger does not pay this number of francs on each kg discharged during the year.

Source: Nicdazo-Crach, J.-L., and Lefrou, C., 1977, *Les Agences Financieres de Bassin*, P. Johanet, Paris, p. 183.

under the household system for paying effluent charges, i.e., payment

of the effluent charge by additional cost per cubic meter of water

intake, or be under the direct dischargers system, i.e., direct payment

to the basin agency on actual or assumed discharge. The determination

is not based on the 6,000 cubic meters per year criterion, but actually

is based on whether or not the discharge contains toxics and/or

whether or not the discharge has a quality corresponding to more than

200 inhabitant-equivalents. If the discharge contains no toxics and

is below the indicated quality limit, there is no direct payment of

effluent charge to the basin agency. In addition, if the activity

meets these limits, but has a water intake greater than 6,000 cubic

meters per year, the activity pays the incremental cost of water

linked to the municipal effluent charge only on the first 6,000 cubic

meters of water intake. Further, such activities are not accounted

for in the computation of the effluent charge of the municipality.

Somehow they are accounted for in the city size (agglomeration)

coefficient. (See below for a description of this last item.)

The base charge, francs per unit of each pollutant discharged

by an activity, is applied in the formula to the discharge on the

normal day in the month of maximum activity. Thus, the total annual

charge for a discharger for a given pollutant is not the unit charge

multiplied by, for example, the total number of kilograms discharged

throughout the year. Rather, the total annual charge is simply what

is calculated in the formula, based on the normal day in the month

of maximum output.[3] Table 12 illustrates the computation of the

effluent charge for 1978 for an hypothetical 100 tons per day bleached

Table 12.   Illustration of Computation of 1978 Effluent Charge on
             Hypothetical 100 Tons Per Day Bleached Kraft Paper Mill
             in the Adour-Garonne Basin

Monthly Production of Mill in 1978			
Month	Production, Tons	Month	Production, Tons
January	2000	July	1900
February	2000	August	500
March	2400	September	2200
April	2500	October	2500
May	2700	November	2400
June	2000	December	2100
Total production in 1978: 25,200 tons			

Month of maximum production: May, 2700 tons

Number of days of production in May: 30 (essentially no downtime for
  maintenance or breakdowns)

VOL = 2700/30 = 90 tons per day (production within May actually
  varied from 55 to 120 tons per day)

KDIS (P) = 8 kg TSS/ton; 36 kg OM/ton; 0.035 kg toxics/ton[a]

KZONE = 1.1

BASE (P) = 34 francs/kg TSS; 46 francs/kg OM; 650 francs/kg toxics

Total effluent charge for the year 1978:

$$90 \ [(8)(34) + (36)(46) + (0.035)(650)][1.1]$$

  TSS   OM   toxics

$$90 \ [272 + 1656 + 22.75][1.1]$$

$$\cong 193 \times 10^3 \ \text{francs} \ [4]$$

[a]These discharges assume removals of 80% for TSS, 60% for OM,
  and 90% for toxics.

kraft paper mill in the Adour-Garonne Basin.

For Municipalities

For municipalities the annual effluent charge is computed according to the following formula:

$$TECHARGE = NBIE \times KSIZE \times KZONE \times BASE,$$

where, TECHARGE = total annual effluent charge for the municipality;

NBIE = number of inhabitant-equivalents discharging into the municipal sewage system;

KSIZE = city size coefficient which reflects that the average pollutant generation per inhabitant-equivalent increases with city size;

KZONE = zone coefficient, the magnitude of which depends on the location of the municipality within the basin; and

BASE = base charge coefficient, francs per inhabitant-equivalent.

The variables KZONE and BASE are identical to those discussed above in relation to activities other than municipalities. The remaining variables are discussed in the following paragraphs.

NBIE. The volume of activity for a municipality is its population. The number of inhabitant-equivalents is calculated as the sum of the resident population plus 0.4 times the estimated seasonal population. The latter is based on the number of hotel rooms, the size of camping grounds, and related factors.

KSIZE. The city size coefficient varies from 0.5 to 1.4, as shown in table 13. The underlying assumption is that the range of

Table 13.   Effluent Charge Coefficients for Urban Centers

Class	Population	Coefficient
I	$\leq$ 500	0.50
II	501 - 2,000	0.75
III	2,001 - 10,000	1.00
IV	10,001 - 50,000	1.10
V	> 50,000	1.20
VI	Paris[a]	1.40
VII	Communes with no water distribution system	-0-

[a]Departements: Paris, Hauts-de-Seine, Seine-Saint-Denis, Val-de-Marne

Source: Arrete (decree) of 28 October 1975, article 17

nondomestic activities increases with city size, and hence the amount of pollutant generation represented by an inhabitant-equivalent increases with city size. (Pollutant generation per inhabitant-equivalent has been set at 90 grams of suspended solids and 57 grams of oxidizable materials per day, a total of 147 grams per day. For comparison, the corresponding values for The Netherlands and the Federal Republic of Germany are both 180 grams per day.) Because small nondomestic activities, e.g., commercial, institutional, small industrial, do not pay effluent charges directly on their discharges, their contribution to the total municipal discharge is purported to be reflected in the city size coefficients.[5]

It bears reiterating, that, as pointed out in the previous chapter, the total annual effluent charge for the municipality is not paid by the municipality, nor by the individual water-using activities in the municipality, to the basin agency. Rather, the effluent charge is paid as an extra cost per cubic meter of water intake to the water distribution company, which then pays the money to the basin agency.

COMPUTATION OF THE PREMIUM
AND THE SUPERPREMIUM

The Premium

Because directly discharging activities other than municipalities pay effluent charges on their actual discharges, premiums are not paid to such activities, but only to operators of municipal treatment plants. The premium is based on the amounts of pollutants, TSS and OM, removed

in treatment.   In effect, the premium equals the effluent charge
multiplied by the amount removed in treatment, except that the coefficient for city size is always 1.0 for the computation of the premium.
To determine the amounts removed, for each treatment plant generally
four measurements are actually made each year by the staff of the basin
agency, or by the department technical assistance team. If done by
the latter, the basin agency contributes toward the costs of measurement.
For large treatment plants, usually the staff of the basin agency
makes all the measurements by itself, including--at least once a year--a
measurement throughout a 24-hour period, with four samples per hour at
regular intervals.

The annual premium paid to the municipal treatment plant operator
is thus calculated in the following manner.

$$\text{Annual PR} = \text{PR(TSS)} + \text{PR(OM)}$$
$$= \text{(kg TSS removed)(BASE TSS)(KSIZE)(KZONE)}$$
$$+ \text{(kg OM removed)(BASE OM)(KSIZE)(KZONE)}$$

where, PR = premium, and the other variables are as defined previously.

The Superpremium

For municipal sewage treatment plants, the annual superpremium
paid to the treatment plant operator is calculated in the following
manner.

$$\text{Annual SPR} = \text{SPR(TSS)} + \text{SPR(OM)}$$
$$= \text{(kg TSS removed)(SPRF TSS)}(C_L)(C_E\text{TSS})$$
$$+ \text{(kg OM removed)(SPRF OM)}(C_L)(C_E\text{OM})$$

where,   SPR   =   superpremium;

  SPRF TSS and
       SPRF OM   =   unit payment factors for TSS and OM removal,
                     respectively, francs per kg removed per day,
                     based on average treatment plant influent;

           C_L =   a coefficient related to the load factor of
                     the treatment plant for the year; and

  C_ETSS and
       C_EOM   =   coefficients related to the average efficiency
                     of the treatment plant in terms of % removal
                     of TSS and OM, respectively.

The superpremium payment factors for TSS and OM are shown in table 14.

Table 14.   Payment Factors Used in Computing Superpremiums
            for Treatment Plants

Average influent to treatment plant in inhabitant-equivalents I	Payment factor, francs per kilo-gram removed per day	
	OM	TSS
$0 < I < 2,000$	156	52
$2,000 < I \leq 10,000$	135	45
$10,000 < I \leq 50,000$	117	39
$50,000 < I \leq 100,000$	108	36
$100,000 < I \leq 800,000$	84	28
$I > 800,000$	$156X^a$	$52X^a$

[a] $X = 0.3 + \dfrac{176,400}{I}$, for 1971 to 1978;

$X = 0.4 + \dfrac{115,000}{I}$, from 1979 to 1982-83

Source: Bulletin de Liaison du Comite de Bassin et de l'Agence Financiere
de Bassin "Seine-Normandie," No. 32, September 1976, p. 58.

The treatment plant load factor (coefficient de charge de la station) equals the average hydraulic load to the plant during the year divided by the nominal capacity of the plant. C_L is then obtained from the relationship shown in figure 12A. The efficiency coefficients for TSS and OM depend on: (1) the type of treatment technology; and (2) how well the treatment facility is operated. They are obtained from measurements of treatment plant influent and effluent, as indicated above. C_ETSS is then obtained from the relationship shown in figure 12B, and C_EOM from figure 12C.

If a directly discharging activity (other than a municipality) has a treatment plant, the activity may receive a superpremium, depending on the performance of its treatment plant. The superpremium is based on the quantities of TSS and OM actually removed, as for a municipal treatment plant. The calculation of the superpremium is made in the following manner.

$$\text{Annual SPR} = \text{SPR(TSS)} + \text{SPR(OM)}$$

$$= \text{(kg TSS removed)(SPRF TSS)}(C_E{}^*\text{TSS})$$

$$+ \text{(kg OM removed)(SPRF OM)}(C_E{}^*\text{OM})$$

where,   SPR   =   superpremium;

SPRF TSS and
SPRF OM   =   unit payment factors for TSS and OM, respectively, as defined above for municipal treatment plants; and

$C_E{}^*$TSS and
$C_E{}^*$OM   =   coefficients related to the average efficiency of the treatment plant in terms of % removal of TSS and OM, respectively.

Figure 12. Relationships Used for Obtaining Coefficients for Use in Computing Superpremiums for Municipal Treatment Plants

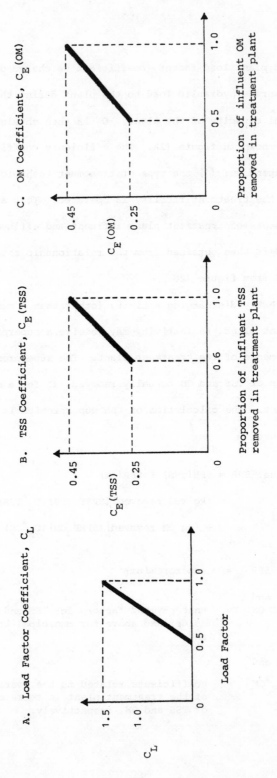

A. Load Factor Coefficient, C_L

B. TSS Coefficient, C_E(TSS)

C. OM Coefficient, C_E(OM)

Source: Bulletin de Liaison du Comite de Bassin et de l'Agence Financiere de Bassin "Seine-Normandie," No. 32, September 1976, p. 60.

The superpremium payment factors for TSS and OM are taken from table 14. The C_E^* coefficients for TSS and OM are obtained from the relationships shown in figures 13A and 13B, respectively.

In situations where ambient water quality objectives exist, a maximum permissible discharge (flux maximum admissible) is computed for each discharger, as described in chapter 6. If the discharge is less than the permissible discharge, the basin agency has no reason to provide an incentive to that discharger for further discharge reduction, because the ambient water quality objectives are reached with the permissible discharge. Therefore, the premium and/or super-premium are calculated as if the discharger were located in the lowest charge zone, regardless of his actual location. This means that it is governmental policy to permit use of the assimilative capacity up to the limit in relation to the ambient water quality objective.

Examples of Calculations
of Premium and Superpremium

The procedures described above are illustrated by applying them to a municipality with an influent of about 150,000 inhabitant-equivalents and the bleached kraft mill of table 12. Both are located in the Adour-Garonne basin in a zone with KZONE equal to 1.1. To highlight the comparison between the municipality and the paper mill, the assumed average influent loads are 3600 kgs TSS and 8100 kgs OM per day and the assumed removal efficiencies are 80% for TSS and 60% for OM, i.e., 2880 kgs and 4860 kgs removed respectively, for both the muni-cipality and the paper mill. The load factor for the municipal treatment plant is 0.7. Values refer to the year 1978.

Figure 13.   Relationships between Removal Efficiency and C_E^*, for
Directly Discharging Activities with Treatment Plants

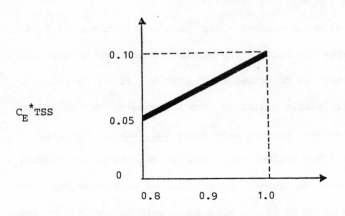

C_E^*TSS

Proportion of influent TSS
removed in treatment plant

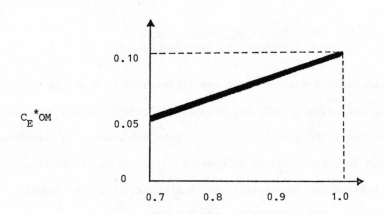

C_E^*OM

Proportion of influent OM
removed in treatment plant

Source: <u>Bulletin de Liaison du Comite de Bassin et de l'Agence
Financiere de Bassin "Seine-Normandie,"</u> No. 32,
September 1976, p. 61.

Municipality

Premium paid by basin agency to treatment plant operator =

2880	x	34	x	1.0	x	1.1
TSS removed		base charge		KSIZE		KZONE

+	4860	x	46	x	1.0	x	1.1
	OM removed		base charge		KSIZE		KZONE

$\cong$ 108×10^3 + 246×10^3 $\cong$ 354×10^3 francs.

Superpremium paid by basin agency to treatment plant operator =

2880	x	28	x	0.7	x	0.35
TSS removed		SPRF TSS from table 14 for influent of about 150,000 I.E.		load factor coefficient		removal efficiency coefficient

+	4860	x	84	x	0.7	x	0.30
	OM removed		SPRF OM from table 14 for influent of about 150,000 I.E.		load factor coefficient		removal efficiency coefficient

$\cong$ 20×10^3 + 86×10^3 $\cong$ 106×10^3 francs

Paper Mill

Superpremium paid by basin agency to paper mill =

$$
\begin{array}{ccccc}
2880 & x & 28 & x & 0.05 \\
\text{TSS} & & \text{SPRF TSS} & & \text{removal} \\
\text{removed} & & & & \text{efficiency} \\
& & & & \text{coefficient}
\end{array}
$$

$$
\begin{array}{ccccc}
+ \quad 4860 & x & 84 & x & 0 \\
\text{OM} & & \text{SPRF OM} & & \text{removal} \\
\text{removed} & & & & \text{efficiency} \\
& & & & \text{coefficient}
\end{array}
$$

$$
\cong \quad 4 \times 10^3 + 0 \quad \cong \quad 4 \times 10^3 \text{ francs.}
$$

The paper mill receives no superpremium for OM removal because the removal efficiency is less than 70%.

The results for the municipality and the paper mill are summarized in table 15.

PERSPECTIVE ON WATER INTAKE CHARGES AND EFFLUENT CHARGES

In 1978, an industrial activity paid, on the average, the equivalent of 0.5% of the value added in his activity in effluent charges. A limit of 3.2% of value added has been placed on the effluent charge imposed on any given discharger.

In 1978, total revenues in France from water intake charges were about 360 million current francs (about 90 million 1978 U. S. dollars). In the same year, total revenues from effluent charges on

Table 15.   Annual Charges, Premiums, and Superpremiums for
             Municipality of 150,000 Inhabitant-Equivalents
             and 100 Tons Per Day Bleached Kraft Paper Mill,
             Adour-Garonne Basin, 1978

Annual Charges, Premiums, and Superpremiums	Municipality		Paper Mill
	Water User	Municipal Government	
Effluent charge paid to basin agency (excluding charge for toxics), 10^3 francs per year[a]	1,105		193
Premium paid by basin agency, 10^3 francs per year		354	0
Superpremium paid by basin agency, 10^3 francs per year		106	4
Net payment, per year	1,105	-460	189

a
 For water users: 150,000 I.E. x 5.58 francs/I.E. x   1.2   x   1.1

$$= 1,105 \times 10^3 \text{ francs;}$$

                                                          KSIZE   KZONE

    for paper mill: from table 12.

net discharges amounted to about 1 billion current francs (about
250 million 1978 U. S. dollars). The total effluent charges paid
in 1978 represented about 20 francs per inhabitant for the year.

NOTES

[1]The sewage tax is also paid as an addition to the water intake
charge.

[2]See the next to last paragraph in this section.

[3]The rate of charge per unit of discharge is totally independent
of where the discharger stands regarding his discharge standards.
The unit charge depends solely on the policy of the basin agency and
the location of the discharger within the basin, e.g., upstream or
downstream.

[4]If the hypothetical paper mill were discharging into a municipal
sewage system, a comparison with typical sewer charges imposed in the
U. S. can be made. The results are informative. Confining the
comparison to total suspended solids (TSS), in the U. S. situation
the mill might pay on every kilogram of TSS discharged into the
sewage system throughout the year. $0.05 per pound of TSS was a
rough average charge in a survey of several U. S. cities in 1976-77.
This translates to about 0.5 francs per kilogram of TSS discharged.
If the mill had discharged 8 kilograms per ton into the U. S. system,
the annual charge on TSS would have been about 100×10^3 francs. This
is in contrast to the about 27×10^3 francs in the French case. In
addition, there would have been a charge on the volume of discharge
in the U. S. case, as well as charges on other pollutants, as in
the French case.

[5]It is not known whether or not there are empirical data to
support the coefficients shown in the table.

Chapter 9

HOW THE LEVELS OF THE EFFLUENT CHARGES ARE SET

INTRODUCTION: BALANCING THE MULTI-YEAR
ACTION PROGRAM AND THE LEVELS OF
THE EFFLUENT CHARGES

Article 14-2 of the 1964 law on water states: "The total amount of charges raised by each (Basin) Agency is determined according to the cost for the Agency of a multi-year action program set in conformity to the orientations of the national socioeconomic plan, and as such, put in an annex to the law which approves it."[1] Therefore, the basic rule in establishing the set of effluent charges--i.e., currently on four substances--is to balance the budget of each basin agency. The problem now is to understand how the multi-year action program itself is established. As described in Section I, there are national long-term objectives at which the multi-year water resources programs, quantity and quality, must be aimed. But within that framework there is a rather large margin of maneuver and the multi-year action programs of the basin agencies are certainly not determined automatically by the long-term national objectives. Not only are there several paths to achieve the long-term objectives, but also, these long-term objectives are not really official, in the sense that they do not result from any official commitment of the government nor from any law. They can always be pushed back a few years, as actually happens. Therefore the problem should be put in the following way: within the process by which

the multi-year action programs and the levels of the effluent charges

are <u>simultaneously</u> set by each basin agency, who are the different

actors, what are their objectives regarding the levels of the charges,

and what is the balance of power among them?

THE PROCESS OF SETTING THE MULTI-YEAR
ACTION PROGRAM AND THE LEVELS OF THE
EFFLUENT CHARGES

The first step in the process is the technical analysis. Given

the estimates of the levels and spatial pattern of activities in the

various coefficient zones of the basin over the five-year planning

horizon, and estimates of the generation and discharge of pollutants

associated with these activities, the assessment basis for the charges

is determined. Then, given the objectives defined by the basin agency,

e.g., degree of treatment to be achieved and the amounts of financial

aids to be provided, the cost to the agency over the five-year period

is obtained. The assessment basis and the costs yield the levels of

charges on the different pollutants. Then the consistency between the

estimated effects of the multi-year action program, in terms of reductions

in discharges, is tested against the initial assumptions concerning the

assessment basis. Several multi-year action programs are developed and

tested in this manner by the staff of the basin agency.

Thus, the basic rule is that there be an overall balance between

revenues and expenses for each agency. Beyond that, there are two

balances which must be respected. One is that there should be a balance

between revenues and expenditures for each pollutant. In fact, one
computation is made for TSS and OM jointly, then one each for toxics
and salts, each of which has generally well identifiable specific costs
of reduction. Separation between TSS and OM is partially arbitrary.
The other balance is for the private sector on one side, and for the
public sector, i.e., municipalities, on the other.

Given these alternative multi-year action programs, the second
step is the choosing among them by the director and the executive board
of the basin agency, to which board the director belongs. Their
choice will reflect their interpretation of the existing balance of
power among the various actors involved in utilization of water
resources in the basin.

The next step involves the Basin Committee and the Delegated
Basin Mission. Article 1 of the 28 October 1975 decree states: "The
Basin Committee is consulted by the president of the executive board
of the basin agency about the levels of the charges to be imposed by
the agency." The Basin Committee, sometimes called the "water
parliament," discusses the charges in relation to the chosen multi-year
action program, as does the Delegated Basin Mission.

The process then moves to the national level, where the Interministerial Mission on Water has the last word. Even though, theoretically,
the Interministerial Mission only discusses the action program, in
reality the discussion covers both program and charges. Negotiations
take place among the ministries involved, and between the Interministerial
Mission and each basin agency. The Water Service of the Ministry of
the Environment acts as the interface between the interministerial level

and the basin agency level, reflecting the national orientations and
the regional orientations, respectively. An example of the process of
computing an effluent charge in relation to a basin agency's multi-year
action program is given in the appendix to this chapter.

ACTORS AND THE
BALANCE OF POWER

From the description of the process by which the levels of the
charges are set, it appears that there are two factors on which the
positions of the actors are based. These are: (1) the institutions
to which they belong; and (2) their national or basin level standing.

National Versus Basin Balance of Power

In the definition of the multi-year action program and of the levels
of the charges, what comes from the basin level has a predominant weight
over purely national considerations. This is understandable because
what comes from the basin level has the weight of: (1) the approval by
a representative assembly, e.g., the Basin Committee; and (2) the approval
by the various concerned ministries through both the Basin Committee,
1/3 of whose members represent the various ministries, and the Delegated
Basin Mission, 100% of its members from the various ministries. However,
the affirmation of the preponderant role of the basin should be moderated
because the starting point of the whole process--that is, the executive
board of the basin agency--is composed so that half of its members are
representatives of the central government. Therefore, one can say
that the central government actually has the first word and the final word.

The Institutional Balance of Power: The Actors

With respect to the setting of the levels of the effluent charges, there are only four actors, besides the basin agency itself, which really carry weight. These are: (1) the technical ministries, e.g., environment, agriculture, industry; (2) the Ministry of Economics and Finance, which, at the national level, has a key role and a veto power; (3) the industries in the basin; and (4) the municipalities in the basin, more generally, individuals locally elected, e.g., members of municipal and general councils. It should be noted that industrial activities and municipalities are present only in the Basin Committee and on the executive board of the basin agency, i.e., at the basin level. However, the ministries are present at every step of the process and at every level. That is, they are: on the executive board of the basin agency--50% of its members, including the director of the agency; the Basin Committee--one-third of its members; the Delegated Basin Mission--100% of its members; and the Interministerial Mission and the Water Service--100% of their members.

The above shows the real measure of the balance of power between the administration at large and the municipalities and industrial activities in the basin. These last two actors necessarily have a reactive role, basically responding to situations and suggestions. However, because these two groups comprise two-thirds of the membership of the Basin Committee, together they have an effective veto power.

However, as it often happens in France, the most important negotiation and discussion take place within the administration itself without any other input, in a non-public way. Therefore what is determinant is the balance of power among the involved ministries. This depends on: (1) their legal attributions and traditional roles, in which respects the Ministry of Economics and Finance is the most important; (2) their possible national-departmental organization and cohesion through a corps of very well trained individuals, from which comes the relative power of the ministries of environment, agriculture, and industry; and (3) the personality of the minister in charge and/or weight given to the individual ministry for general political reasons, by the prime minister and the president of the country.

The Technical Ministries

On the whole, for the specific matter of setting the levels of charges, the technical ministries have similar attitudes. They tend to support a significant increase in the charges in order to achieve the national objectives. This is also the position of the directors of the basin agencies. Besides their legal attributions, these institutions have an additional share of the power because they are the only ones who have the technical knowledge about what is discussed, i.e., the multi-year action program for water management.

Furthermore, the director of the agency is the one who proposes, i.e., who takes the first step. Discussion is initiated by him through his proposals and he can argue on technical grounds. This gives him a definite advantage over industrial activities and municipalities and allows him to have a good balance of power vis à vis the Ministry of Economics and Finance. Generally the director would favor a rather substantial increase in the charges.

The Ministry of Economics and Finance

The Ministry of Economics and Finance is the most influential
of the ministerial actors and generally favors a much slower increase
in the charges than the other ministries for two independent and
important reasons. First, the ministry considers that charges are a
non-productive use of the scarce resources of industry, which should
rather be invested to compete more favorably in foreign markets.
In addition, the effluent charges are considered by the ministry to
be very inflationary. Second, as noted previously, effluent charges
and water intake charges represent a very special sort of charge,
one which is uniquely not instituted by the Ministry of Economics and
Finance and therefore is outside of its direct control. This circum-
stance is an important reason why this ministry has had a constant
policy of minimizing these charges. Another example: all efforts
made since 1964 to institute charges on air pollution emissions and
on noise emissions--the latter especially by aircraft--have been strongly
opposed by the Ministry of Economics and Finance. Given its
enormous real power, this Ministry has had no difficulty in preventing
imposition of such charges.

A good illustration of the power and of the veto power of the
Ministry of Economics and Finances occurred in 1976 when the anti-
inflationary "Barre plan" was applied. Even though the multi-
year action programs of the basin agencies had been agreed upon before
the plan was promulgated and the levels of charges had been implicitly
agreed upon for the next few years, the Ministry of Economics and

Finance demanded and obtained an increase of only 6.5% in the charges

for 1977 and 1978. (Mr. Raymond Barre headed the ministry at that

time, as well as being prime minister.) In constant francs this

increase actually represented a decrease of about 4% in the value

of the charge. This decision disrupted the financial equilibrium

of the basin agencies and compelled modifications of all of their

multi-year action programs.

Industrial Activities and
the Municipalities

These two actors have roughly the same importance. Interestingly

enough, since the inception of the effluent charges system, industry

has been a more cooperative partner with respect to the evolution of

the levels of effluent charges than would have been expected a priori.

On the whole, industry has accepted fairly well the role of the basin

agencies and the payment of increasing effluent charges.[2] One of the

reasons for this attitude might be that industrial entrepreneurs

prefer to pay a rather moderate amount of money, some of which they can

recover if needed through aid, than to be confronted directly by the

either too absolute or non-understandable effluent standards. The

basin agency, which speaks both the technical and the financial

language, is probably, for an industrialist, an easier institution to

which to relate and with which to get along than the departmental

administration which speaks the administrative and regulatory

language. Note that generally speaking, industry has little

or no direct influence on governmental administration, even

though some major groups have their direct lines of action and
pressure. Industry and government really are two different worlds.

Municipalities, on the contrary, have sometimes been a difficult
partner for the basin agencies. An example of that is the previously
mentioned legal battle of the Association of the Mayors of France
against having to pay the charges. The reason for the opposition is
the permanent financial problems of the small municipalities which
apply for departmental, regional, and national grants for almost
everything they try to do. The effluent charges had been perceived
as a new tax which was not offset by any new revenue. It now appears
that the vast majority of the municipalities is taking a more positive
approach to increasing the effluent charges, especially because the
municipalities do not pay the charges directly from their budgets.

REACHING AN EQUILIBRIUM

Having made a first estimate about the political and economic
contexts at the national and basin levels, the director of the basin
agency formulates an hypothesis about what levels of charges he can expect to
be approved. He then builds his draft of the multi-year action program
on that basis, and subsequently defines his tactics to argue and to
obtain those levels of charges necessary to implement the action program.

Then comes the establishing of a balance within the administration,
represented by the technical ministries on one side and the Ministry
of Economics and Finances on the other. This confrontation sets the
range of values for the effluent charges that will be considered

reasonable. Then the industrial activities and municipalities in the
basin select the equilibrium levels, generally at some intermediate
levels on which all the actors will finally agree.

In fact, this search for an equilibrium can well take place in
an informal manner before a detailed action program is designed. It
can also happen that, after an action program and related set of
charges have been determined, a powerful actor--such as the Ministry
of Economics and Finance--changes its mind and decides to shift the
equilibrium. (This is most likely when the minister of economics and
finance is also the prime minister.)

Basically, reaching an equilibrium involves an iterative process.
Either one or both of two results will require iteration. One, if the proposed program of the basin agency results in levels of charges which
are politically unacceptable, the program must be revised. Two, if
the levels of charges associated with the program of the basin agency
are likely to induce more reductions in discharges than anticipated,
then revenues from the charges would be insufficient to cover the
proposed expenditures. The appendix to this chapter illustrates the
relationship between the program and effluent charges of a basin agency.

Figure 14 is an attempt to depict the process described in the
preceding paragraphs.[3] Any attempt to depict such a complicated
process can only be sketchy. However, the objective of the effort
is to try to characterize the game and the interplay of the actors in
that game, so that the famous decision-making process becomes something
more than the traditional black box, which constitutes a non-acceptable

description for the analyst. The assumption herein is that any try

at shedding some light on the process, as imperfect as it may be,

is better than ignoring the question.

NOTES

[1]The multi-year action program is more appropriately termed a multi-year program of interventions, because a basin agency is limited to providing financial aid and technical advice, and making studies and plans.

[2]This of course may have been because the levels of charges were relatively low, and because of the significant subsidies the charges made possible.

[3]As mentioned above, the process is done three times: once for SS and OM, then for toxics, and then for salts. The process is essentially the same for establishing water intake (abstraction) charges, i.e., the quantity side of water resources management.

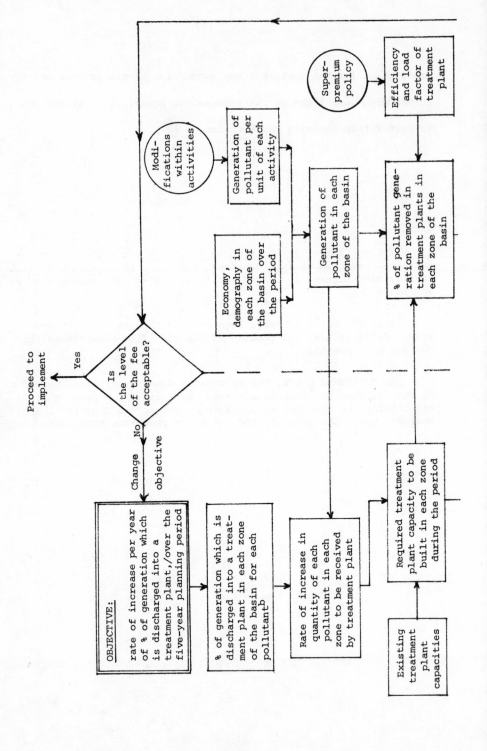

Figure 14. Basin Agency Procedure for Computation of Level of Effluent Charge and the Interrelationship between the Charge and the Objective[a]

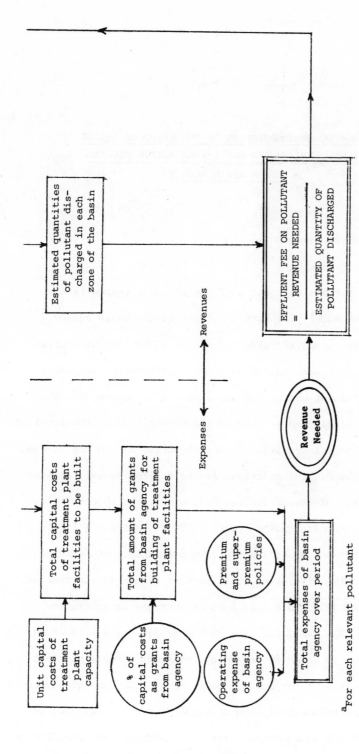

Unit capital costs of treatment plant capacity

Total capital costs of treatment plant facilities to be built

% of capital costs as grants from basin agency

Total amount of grants from basin agency for building of treatment plant facilities

Premium and super-premium policies

Operating expense of basin agency

Total expenses of basin agency over period

Revenue Needed

Expenses

Revenues

Estimated quantities of pollutant dis-charged in each zone of the basin

EFFLUENT FEE ON POLLUTANT = $\dfrac{\text{REVENUE NEEDED}}{\text{ESTIMATED QUANTITY OF POLLUTANT DISCHARGED}}$

[a]For each relevant pollutant

[b]Based on: estimated generation in relation to estimated levels of population and activities; assumed changes in technology, product mix, product specifications; assumed O&M procedures

Indicates function of policy of basin agency

Appendix to Chapter 9

EXAMPLE OF COMPUTATION OF THE EFFLUENT CHARGE
IN RELATION TO THE MULTI-YEAR ACTION PROGRAM
OF A BASIN AGENCY

CONTEXT OF AND
ASSUMPTIONS FOR THE EXAMPLE

The basis for the unit level of the effluent charge is the
balancing of the multi-year action program of the basin agency.
Therefore, there is an iterative process of adjusting the desired
water quality management objective with a politically acceptable
level of the effluent charge, as was depicted in figure 14.   The
following is a simplified example of the computation of the level of
the effluent charge in relation to the objective and to the various
parameters involved.   The objective is the desired annual increase
in the percentage of generation treated, where this objective is
defined only in terms of point sources.   Thus, CHARGE = f (OBJ),
where, CHARGE = level of effluent charge per inhabitant-equivalent;
and OBJ = objective, for given year, percent of generation treated.
Between 1969 and 1977 the annual rate of increase in generation
treated, i.e., OBJ, was 5.5%.   For 1977 and 1978 it was about 2%.

The example is based on an average municipal treatment plant
in the average zone of the basin.   Only the equilibrium for suspended
solids (TSS) and biochemical oxygen demand (BOD_5) is included, in
terms of inhabitant-equivalents, without considering the question of

the load factor of the treatment plant. All expenses other than
for sewage treatment plants are computed as a percentage of the
treatment plant costs. All values correspond approximately to the
1978 situation in the Seine-Normandie Basin.

Table 16 defines the parameters involved in the computation
of the charge. The values of these parameters shown in the table
are the 1975-1980 values.

COMPUTATION OF THE BUDGET
AND THE CHARGE

<u>Expenses</u>

The expenses of the basin agency for 1978 are:

Premiums for treatment = assessment basis x unit charge

= (GEN)(PGT)(EFF) x (CHARGE);

Superpremiums for treatment

= (GEN)(PGT)(EFF) x (CHARGE) x (SUPR);

Subsidies for capital costs of treatment

= (GEN)(dVGT)(KT) x (KAID)

Subsidies for other purposes

= (GEN)(dVGT)(KT)(KAID) x (ZAID); and

Operating expenses of agency = (AOMC) x (total expenditures).

Therefore, total expenses =

$$\frac{GEN\left\{[(PGT)(EFF)(CHARGE)][1 + SUPR] + [(dVGT)(KT)(KAID)][1 + ZAID]\right\}}{(1 - AOMC)}$$

Table 16. Parameters and Parameter Values Used in Computation of the Effluent Charge for 1978, Seine-Normandie Basin Agency

Symbol	Definition of Parameter Symbol	Factors Affecting Parameter Value	Value	Comments
KT	Total capital cost for an average treatment plant per inhabitant-equivalent, e.g., 90 grams of SS, 57 grams of BOD_5 per day per year, for specified efficiency of removal	Depends on the state of technology, productivity of the economy, and relative prices	200 francs	Represents total capital cost in 1978 francs, not an annual cost
EFF	Average efficiency of treatment, percent removal of influent load	Depends on the operating and maintenance efficiency of treatment plants, the incentive power of the superpremium, and the amount of technical assistance	70%	Efficiency was about 60% in 1970. It steadily increased since then because of the superpremium and technical assistance policies, and reached about 68% in 1978.
PGT	Percentage of generation by point sources which is influent to a treatment plant	Depends on the objective; it is the only parameter which changes even in a steady-state situation	55%	The percentage of generation treated increased from 31% in 1969 to 51% in 1977 and to 55% in 1978.
OBJ	Annual rate of increase in the percentage of generation which is influent to a treatment plant	This parameter represents the objective of water quality management. Depends on rate of construction of treatment plants and sewage collection networks	2%	Between 1969 and 1977 this rate was about 5.5%; for 1977 and 1978 it was about 2%.
GEN	Generation in year 1978	Depends on population levels and activities	--	--

182

dVGT	Rate of increase of generation which is influent to a treatment plant	--	6%	The rate of increase of generation treated is the sum of: (1) the rate of increase in generation; and (2) the rate of increase in % treated. For 1978 these values are 4% and 2%, respectively.
KAID	Average percentage aid for capital costs of sewage treatment plant construction	Depends on the policy of the multi-year action program	30%	--
ZAID	Value of all other subsidies by and expenses of the basin agency, expressed as % of the subsidies for sewage treatment plants	Depends on the policy of the multi-year action program	50%	Includes subsidies for: sewage collection systems; facilities and premium for reducing discharges of toxics; technical assistance; measurements; and studies
SUPR	Average value of the super-premium, expressed as % of the premium	Depends on the policy of the basin agency	20%	--
AOMC	Operation and maintenance costs of the basin agency, expressed as a % of its total expenditures	Depends on the role and activities of the basin agency	7%	--
CHARGE	Level of the effluent charge per inhabitant-equivalent	To be calculated	--	Level of charge is to be sufficient to insure equilibrium of the budget of the basin agency

Revenues

The revenues of the agency for 1978 equal the assessment basis, GEN, multiplied by the unit charge for the given pollutant, CHARGE per inhabitant-equivalent, i.e., total revenue = GEN x CHARGE.

The Charge

Because total revenues must equal total expenses, then,

$$CHARGE = \frac{[(dVGT)(KT)(KAID)][1 + ZAID]}{(1 - AOMC) - (1 + SUPR)(PGT)(EFF)}$$

For the 1978 values of the parameters, the resulting base charge is about ten francs per inhabitant-equivalent, which is the level of charge valid in 1978 for the Seine-Normandie Basin Agency. (The application of the formula actually yields 11.4 francs, which, given the zone and agglomeration coefficients in the basin, is the average charge paid per inhabitant-equivalent in 1978.)

APPLICATION IN RELATION TO
1990 ALTERNATIVE OBJECTIVES

Table 17 shows parameter values associated with the desirable and minimum objectives for 1990. Values of parameters in the charge formula not listed are the same as in table 16. Generation is estimated to increase at a rate of 4% per year over the 1978-1990 period. To move in twelve years from 55% to 90% of generation treated, the desirable objective, requires an average increase of 4.2% per

Table 17.   Parameter Values in Relation to 1990 Desirable and
             Minimum Objectives, Seine-Normandie Basin Agency

Parameter	Value of Parameter			Comments
	at 1978	at 1990	Average 1978-1990	
To Reach the 1990 Desirable Objective				
EFF	70%	90%	80%	Assumes that policies with respect to efficiency of treatment are achieved
PGT	55%	90%	73%	Assumes not only that required treatment plants are built, but also the necessary related sewage collection networks; requires 4.2% per year increase
dVGT	--	--	8.2%	
To Reach the 1990 Minimum Objective				
EFF	70%	90%	80%	Assumes that policies with respect to efficiency of treatment are achieved
PGT	55%	70%	63%	Assumes not only that required treatment plants are built, but also the necessary related sewage collection networks; requires 2.0% per year increase
dVGT	--	--	6.0%	

year in the percentage of generation treated. To move in twelve
years from 55% to 70% of generation treated, the minimum objective,
requires an average increase of 2% per year in the percentage of
generation treated. Combining the increase in generation and the
increase in percentage of generation treated yields the values shown
for dVGT of 8.2% and 6.0% for the 1990 desirable and minimum objectives,
respectively.

Applying the formula for the charge, the equilibrium results
(balanced budget) are:

- effluent charge to achieve 1990 <u>desirable</u> objective
  is 32 francs per inhabitant-equivalent; and
- effluent charge to achieve 1990 <u>minimum</u> objective is
  17 francs per inhabitant-equivalent.

COMMENTS

The above demonstrates the computational procedure; the numbers
indicate orders of magnitude. The following conclusions can be drawn.

(1) The 1990 <u>desired</u> objective cannot be reached by 1990,
but rather only by about 2000 at best, because the required evolution
of the effluent charge--20% per year in real value--to achieve that
objective is politically infeasible.

(2) The 1990 <u>minimum</u> objective can be reached provided
that: (a) the level of the charge increases by 6% per year in real
value, which is quite probable; and (b) the average load factor of
the treatment plants does not decrease, which is much less easy to achieve.

(3) The level of the effluent charge corresponding to the 1990 desirable objective, 32 francs, probably is still not sufficient to provide an incentive for municipalities and for industrial activities discharging directly to water bodies, to reduce discharges by installation and operation of conventional treatment plants. The issue of the "incentive power" of the effluent charges is analyzed in the next chapter.

Chapter 10

THE INCENTIVE POWER OF THE EFFLUENT CHARGES

GENERAL PERSPECTIVE

As indicated previously, the word "incentive" itself does not appear
in the legal texts relating to the effluent charges system, nor in the
associated rules, procedures, explanations. For the legislator, the charges
are exclusively dedicated to balancing the budgets of the basin agencies.
They are always seen as a payment for services or as compensations, never
as a tool which can have an impact on demand and on the behavior of dischargers. Nevertheless, almost every time people responsible for water
management in France discuss effluent charges they speak of the incentive
question. For example: "The will of the legislator in 1964 to open...an
economic way for water management...trying to influence water demand
through economic incentives,"[1] and "The effluent charges have a role of
economic incentive and therefore contribute to give a price to water."[2]
It should be emphasized however, that in every case the notion of economic
incentive appears _after_ mention of the regulatory system, to stress the
dual nature of the water management system. Regardless of the words and
of the original objective in establishing the effluent charges system,
the question is: what can be said at this point in time, 1979, regarding
the incentive power of the effluent charges?

ATTEMPTING AN EMPIRICAL ASSESSMENT OF THE
INCENTIVE POWER OF EFFLUENT CHARGES IN FRANCE

Table 18 shows, for 1978 within the area of the Seine-Normandie
Basin Agency, estimated capital and operation and maintenance costs
for a primary plus secondary municipal treatment plant, per year,
per inhabitant-equivalent. The table also shows the effluent charge
per inhabitant-equivalent and the annual premium and superpremium
for the plant.

Table 18. Annual Costs of Primary Plus Secondary Wastewater
Treatment, Typical Municipal Treatment Plant, and
Magnitudes of Premium, Superpremium, and Effluent
Charge, Per Inhabitant-Equivalent, Area of Seine-
Normandie Basin Agency, 1978

Item	Annual Amounts	
	Francs	U.S. $ [a]
Amortization of total capital investment[b]	10.00	2.33
Annual operation and maintenance	20.00	4.65
Portion of annual amortization paid by basin agency (30%)	3.00	0.70
Premium received for treatment[c]	7.00	1.63
Superpremium received for efficiency of plant operation[c]	1.50	0.35
Total Subsidy	11.50	2.68
Effluent charge, francs per I.E.	10.00	2.33

[a]Based on exchange rate of 4.30 francs per U.S. dollar

[b]Assuming 20 years straight line amortization

[c]Assuming 70% removal efficiency

Based on the above data, the following comments on the incentive question can be made. First, it is almost three times as expensive for the municipality to go to secondary treatment as to do nothing. Annual costs of secondary treatment per inhabitant-equivalent of 30 francs, plus the effluent charge on generation of 10 francs, yield a total of 40 francs. Receipts are: the portion of the amortization from the basin agency, 3 francs; the premium, 7 francs; and the superpremium, 1.5 francs for a total of 11.5 francs. The net cost is therefore 28.5 francs per inhabitant-equivalent. Second, the operation and maintenance (O&M) costs of the secondary plant, 20 francs, are larger than the returns from the basin agency in the form of the premium and superpremium, which together total 8.5 francs. Even if it were considered that the treatment plant had been built and the capital costs fully amortized, rational economic behavior would be not to operate the plant and instead pay the effluent charge of 10 francs. If the plant were operated, the payments would be 20 francs for O&M plus 10 francs for the effluent charge (because the charge is on generation), while the receipts would be 8.5 francs, leaving a net cost of 21.5 francs per inhabitant-equivalent.

The foregoing simple discussion belies the more complicated real situation. Therefore, to make a more accurate assessment of the incentive power of the effluent charges, more detailed analysis is necessary by type of activity, i.e., those activities discharging into a municipal sewage system and those activities, mainly industrial, discharging directly to water bodies.

The Incentive of Effluent Charges for
Activities Linked to a Municipal Sewage System

The discussion in this   section is limited to activities with
water intakes less than or equal to 6,000 cubic meters per year and
connected to a municipal sewage system. (Activities connected to
a municipal sewage system but with intakes greater than 6,000 cubic
meters per year or meeting certain other criteria are treated the
same as activities discharging directly to water bodies, and are
discussed in the next section.) For these activities, the effluent
charge provides virtually no incentive to reduce discharges, because
the effluent charge represents at the most about 5% of the final
total cost of intake water which is effectively billed. Even a
sizeable increase in the charge would have a very small effect on
intake water price, and thus on water intake and therefore on wastewater discharge. In any case, the effect would be much smaller than
the price differences in intake water among municipalities because
of different local conditions. It is important to note that the
dischargers in this category, for whom the effluent charge provides
virtually no incentive to reduce discharges, are responsible for
about 40% of the discharge of BOD_5 and SS (suspended solids) in France.

Turning from the individual activities to the operator of the
sewage treament plant--usually a municipality or a group of munici-
palities--it must be understood that the operator of the treatment
plant does not perceive the potential incentive effect of the effluent
charge directly. This is because the effluent charge is not paid
by the operator of the municipal treatment plant, but rather through

the water distribution company via intake water bills. Nevertheless, even if indirectly through the political process, the level of the effluent charge is conveyed to the treatment plant operator and can affect his behavior regarding discharges, through the attitudes and perceptions transmitted by the political process in the municipality(ies).

It is important to note, in attempting to assess the incentive power of the effluent charge on a municipal sewage treatment plant operator, that amortization of the capital cost of the plant needs virtually no consideration in the analysis. This is true because most of the capital costs are provided by agencies other than the municipality, namely: (1) 30% as outright grants from the basin agency; (2) 20% to 50% as outright grants from departmental, regional, and national sources; and (3) the remaining portion via loans from the Caisse de Depots at very low interest rates, much less than the market rate. Furthermore, municipal officials are elected for only six years, and consequently they are generally not too concerned about commitments for long-term debts.

The Incentive of Effluent Charges
for Activities Discharging Directly

The typical case of this type is an industrial activity which has its own treatment plant. In this case capital costs must be taken into account. Usually the investment will be considered sound only if it can be paid off in no more than five years. The charge will provide an incentive to reduce discharge up to a level where the marginal cost of reducing discharge equals the level of the charge. The incentive power of the charge at its present (1979) level,

i.e., 10 francs per inhabitant-equivalent in the Seine-Normandie
Basin, can be assessed by computing the marginal costs of reducing
discharge through conventional treatment plants—primary, secondary,
and tertiary.

The cost to the discharger is made up of four items, each
expressed in francs per inhabitant-equivalent (I.E.).

(1) The annual capital cost of the treatment plant which the
    discharger has to pay. This is equal to:

$$\frac{KCOST\ (1-c)}{d}$$

where, KCOST = total capital cost of treatment
                facility;

       c = capital grant aid, in percent of
           total capital cost; and

       d = number of years for repayment of
           investment.

(2) The annual operation and maintenance costs, i.e., O&M Cost.

(3) The superpremium paid by the basin agency to the discharger,
    as an aid for O&M Cost. This is a negative cost and is
    equal to:

$$SUPR = (b)(a)(CHARGE)$$

where, SUPR = superpremium;

       b = superpremium rate, %;

       a = degree of removal in treatment, %; and

       CHARGE = level of charge imposed on discharge,
                i.e., francs per unit of pollutant discharged.

(4) The effluent charge. This is equal to:

Effluent charge = (1-a) CHARGE

Therefore, the total cost to the discharger is: (1) + (2) - (3) + (4).
Table 19 shows typical 1979 values for the variables indicated above,
in the Seine-Normandie Basin. Table 20 shows the computation of the
marginal costs of discharge reduction for the discharger, given the
costs and parameter values in table 19.

Table 19. Typical Unit Costs of Three Levels of Conventional
           Treatment for a Directly Discharging Activity, and
           the Level of the Charge, Seine-Normandie Basin

1979 Values	Level of Treatment (I.E.)		
	Primary treatment	Secondary treatment	Tertiary treatment
KCOST, francs per I.E.	80.0	200.0	300.0
O&MCOST, francs per I.E.	6.7	20.0	32.0
a, %	40.0	80.0	93.0
b, %	-0-	20.0	40.0
c, %	30.0	30.0	40.0
d, years	5	5	5
CHARGE, francs per I.E.	10.0	10.0	10.0

KCOST   = total capital costs of treatment facilities

O&MCOST = annual operation and maintenance costs

a   = % removal in treatment

b   = superpremium rate, %

c   = capital grant aid, %

d   = number of years for repayment of investment

Table 20.   Annual Costs and Marginal Costs of Discharge
             Reduction by Means of Conventional Treatment,
             and Corresponding Reductions in Discharge Achieved[a]

Item	Level of Treatment			
	No Treatment	Primary	Secondary	Tertiary
(1) Annual capital cost, francs per I.E.	0	11.2	28.0	36.0
(2) Annual O&M cost, francs per I.E.	0	6.7	20.0	32.0
(3) Annual superpremium (aid)	0	0	1.6	3.7
Total annual cost (exclusive of charge) = (1) + (2) − (3)	0	17.9	46.4	64.3
Units discharged, I.E.	1	0.6	0.2	0.07
Incremental costs of discharge reduction, francs		17.9	28.5	17.9
Incremental discharge reduction, I.E.		0.4	0.4	0.13
Incremental cost per unit of discharge reduced, francs per I.E.		44.8	71.3	137.7
Actual 1979 level of the charge, francs per I.E.		10.0	10.0	10.0

[a]Based on costs and parameter values in table 19.

As stated previously, a rational, i.e., cost minimizing, discharger would reduce his discharge up to the point where his marginal cost of discharge reduction is equal to the level of the charge. Table 20 shows the marginal costs for primary, secondary, and tertiary treatment. These costs mean that the current level of the charge, 10 francs per I.E., should be multiplied by about 4.5, 7, and 14 in order to provide incentive for discharge reduction equivalent to primary, secondary, and tertiary treatment, respectively.

The foregoing calculations do not mean that the effluent charge may not be an incentive to reduce discharge by other means, such as in-plant modifications and recycling. In fact, in industries such as slaughterhouses, pulp and paper manufacturing, metal finishing, and stone cutting, where some reduction in discharge is either possible at low cost or which allow for recycling of useful nonproduct outputs, even the low effluent charge is, to some extent, an incentive. This is even more true for industrial activities discharging to municipal sewage systems, because the sewage tax can be significantly reduced by recycling of water, thereby reducing water intake on which the sewage tax is levied. As noted previously, the sewage tax is, in general, much higher than the effluent charge.

The results shown in table 20 shed light on three other points. These points are valid even though the marginal costs of treatment have been computed and extrapolated from conventional treatment plant technologies, and there are likely to be somewhat cheaper ways to achieve discharge reduction, e.g., through internal recyclying measures. First, discharge standards for activities which have

explicit discharge standards are usually set at levels far beyond
where the level of the effluent charge would naturally drive the
activity to go, in terms of discharge reduction.

Second, the various subsidies for capital costs and O&M costs
do not succeed in giving much incentive power to the charge. That
is, even though these subsidies reduce the costs to the discharger,
as the discharger sees them, the costs to the discharger are still
very substantially higher than the effluent charge. In fact, the
basin agencies are having some difficulty in finding activities
to accept their grants. If the subsidies were removed, the level
of the charge would be even less of an incentive, in relation to
the marginal costs of discharge reduction. This is shown in table 21,
using the same costs and parameter values as used in table 20.[3]

Third, the results show that the effluent charge, at present,
is primarily an instrument for raising revenues rather than an
incentive to induce discharge reduction.

CHARGES, INCENTIVES, AND
OVERALL WATER MANAGEMENT

Given the above assessment, under what rationale are treatment
plants being built and operated in France? How has the gap been
filled between the incentive power of the actual levels of effluent
charges and the levels of charges theoretically needed to induce the
rational decision-maker to initiate action? The answer is by the
regulatory side of the water management system: the gap not covered
by the effluent charges is filled by the complementary incentive

Table 21.   Annual Costs and Marginal Costs of Discharge Reduction
by Means of Conventional Treatment, Without Subsidies[a]

Item	Level of Treatment			
	No Treatment	Primary	Secondary	Tertiary
(1) Annual capital cost, francs per I.E.	0	16.0	40.0	60.0
(2) Annual O&M cost, francs per I.E.	0	6.7	20.0	32.0
(3) Annual superpremium (aid)	0	0	0	0
Total annual cost (exclusive of charge) = (1) + (2) - (3)	0	22.7	60.0	92.0
Units discharged, I.E.	1	0.6	0.2	0.07
Incremental costs of discharge reduction, francs		22.7	37.3	32.0
Incremental discharge reduction, I.E.		0.4	0.4	0.13
Incremental cost per unit of discharge reduced, francs per I.E.		56.8	93.3	246.0
Actual 1979 level of the charge, francs per I.E.		10.0	10.0	10.0

[a]Based on costs and parameter values in table 19.

power of regulations and by the various subsidies.

The situation can be summarized as follows.

● Municipalities are reluctant to act and are not very
susceptible to direct regulations because no ultimate
sanctions can be imposed. Therefore, to induce municipalities to act requires sizeable subsidies for both
capital and operation and maintenance costs.

● Industrial activities are reluctant to act, but are
susceptible to direct regulations because the ultimate
sanctions can be and have been imposed. Therefore lower
subsidies for capital and operation and maintenance
costs are provided than for municipalities. But basically,
industrial activities are moved to act because of the
regulations and because of the available subsidies, not
because of the effluent charges.

If it were not for the "fear of the policeman," not many individual
dischargers would accept the offers of the basin agencies to help
finance their building of treatment plants. If it were not for the
opportunity of a fair bargain with the basin agencies, not many
dischargers would take the discharge standards seriously.[4]

NOTES

[1]Statement of Mr. J. F. Saglio, Director of the Direction of the
Prevention of Pollution, in the preface to Les Agences Financieres
de Bassin, Paris, 1977, p. 3.

[2]*Livre Blanc de l'Eau en France*, Paris, 1974.

[3]Both tables 20 and 21 exclude any consideration of available investment tax credit and rapid depreciation on pollution control equipment.

[4]It is a widely known secret that the regulations regarding effluents, before 1969, were generally disregarded, or the standards were artificially kept out of date through "old" prefectoral decrees.

Chapter 11

FROM A REDISTRIBUTIVE TO AN INCENTIVE
CHARGE SYSTEM: A LONG-TERM STRATEGY TO ENFORCE REGULATIONS

Before indicating where the discussion in the previous chapters
has led in terms of characterizing water quality management in France,
a few salient points are presented for emphasis.

● Water quality management in France involves basically two
sets of institutions, one on the regulatory side and one on the economic
incentive side. Although there are no nationally defined discharge
standards, the Classified Establishments Service does define standards of
generation and discharge per unit of activity for many different types of
activities. These form the basis for the discharge permits issued by the
prefects. Inspections of activities and their performance are made by
inspectors of the Classified Establishments Service, on the basis of which
recommendations for changes, including plant closings, are made. But the
Classified Establishments Service has no funds for subsidies, does not
monitor ambient water quality, does not prepare water quality management
plans. On the other side, the basin agencies: cannot forbid or impose any
actions; cannot have direct relationships with prefectoral actions; are not
responsible for checking whether or not regulatory conditions are achieved;
cannot enforce any law; cannot commission, build, operate facilities
such as treatment plants and reservoirs. But the basin agencies can:
provide financial assistance in terms of grants for capital investments,

grants for operation and maintenance costs, loans for capital investments; check the efficiency of operation of treatment plants; make technical studies and plans; provide technical advice to dischargers; and monitor ambient water quality.

● Dischargers must meet various conditions before subsidies, e.g., grants, are provided. For example, the applicant for a grant must have an official authorization to dispose of sludge in an appropriate place and manner before the subsidy is granted.[1] In some cases sludge treatment or incineration is partially subsidized by the basin agency, but only if it is part of the overall wastewater treatment program.

● The basin agencies can provide subsidies for in-plant changes to reduce discharges, as well as for end-of-pipe treatment plants. Criteria for doing so vary among the basin agencies. One approach ties the extent of financial aid for a treatment plant to the degree to which reasonable in-plant measures to reduce residuals generation have been taken. There also can be aid even if only in-plant measures are taken. The problem then is to determine the basis on which to grant the financial aid.[2]

● Effluent charges are imposed by the basin agencies on discharges of oxidizable material, suspended material, and toxics, and abstraction charges are imposed on water withdrawals. Clearly, the objective of the charges is to raise revenues to cover the expenses of the basin agencies. The revenues raised must be sufficient to balance water quantity and water quality budgets, and municipal and private sector budgets. The charges are not designed as incentives to reduce residuals discharges or water intakes. Depending on the cost functions of individual water uses, the charges may provide incentives to reduce wastewater discharges and/or water intake for some individual activities.

● Neither regulations nor charges have thus far been imposed on nonpoint sources of pollution. Nor does it appear that subsidies have been granted for either capital or operation and maintenance costs to reduce discharges from nonpoint sources.

● Increasingly water quality management in France is being directed toward area, river reach, and aquifer specific ambient water quality objectives, as reflected in the Ambient Water Quality Objectives Policy. To help achieve the objectives of that policy, effluent charges are increased where ambient water quality objectives are <u>not</u> being achieved, and decreased--or kept constant--where the objectives are being met. Similarly, the subsidy--in terms of percent of capital costs of discharge reduction facilities provided--is increased in areas where the ambient water quality objectives are not being met.

WATER QUALITY MANAGEMENT IN FRANCE:
EVOLUTION, PRESENT STATUS, PROSPECTS

The regulatory apparatus for water quality management in France, when it was not complemented by the effluent charges system, led to a sort of dead end. There were three reasons for this. One, the regulations were often impossible to enforce because the system appeared both inequitable and inefficient. The regulations were often <u>inequitable</u> among dischargers because prefectoral decrees--the bases for the regulations--reflect a local balance of political power rather than the local assimilative capacity of the water environment. The

regulations were <u>inefficient</u> because they could not take into account

local conditions, and it was perceived as very inefficient to treat

a waste discharge to a level cleaner than the river to which it is

discharged. It was also inefficient to fight with small, old plants

which were unable to meet <u>any</u> standards. Two, adequate enforcement

would have required a much larger and better trained corps of inspectors

than was available, whose major role would have been to impose

sanctions for noncompliance with the regulations. Three, municipal

wastewater discharges were essentially unaffected by regulations,

because the problem of reducing such discharges is basically financial.

The foregoing means that there was no politically acceptable

way to get out of the dead end along the line of better enforcement

of the existing tools, i.e., standards and regulations. The system

needed both a bargaining table--the Basin Committee--and a bargaining

tool--grants and loans. It desperately needed some room and some "gas"

to function. However, to start movement, there had to be a genuine

strategy about the level of the effluent charge, which had to be

accepted by all parties. The path was narrow because, at the beginning,

in case of major conflict between the agencies and the dischargers,

the whole idea of charges could well have been questioned.[3] The

acceptance of the effluent charges system has been achieved by:

(1) setting a flat rate assessment basis for all pollutants, which

level was obviously too low to induce, by itself, reductions in dis-

charges; and (2) beginning with low rates for the charges, which were

about 15 times below an incentive level in 1969, and gradually increasing

the charge rates. In fact, since the beginning, the agencies have
had no significant problems in having the charges paid, except for
the controversy with the municipalities.

The effluent charges system then does not have much to do with
what theory says it should be, and even not much to do with the Polluter-
Pays Principle. But the effluent charges system, as unsatisfactory
and even strange as it looks, bears only a difference of degree and not
of nature with something satisfactory, i.e., a system which leads to
the objective of cleaning up water bodies while respecting the Polluter-
Pays Principle. For this to be the case required that the system have
two characteristics: (1) the discharges be effectively measured, to
preclude having too low an assessment basis; and (2) the rate of charge
be significantly increased over time. These two have effectively been
consistent lines of action, and even combat, on the part of the basin
agencies. One needed first to initiate the system, even at the price of
an unsatisfactory system, as long as the inadequacy came from a matter
of degree--level of charge on a given parameter--and not from the basic
principle. Now, in 1979, the situation is somewhere between this
initial state of 1969 and a fully appropriate effluent charges system.
The levels of the charges are still not much of an incentive to reduce
discharges and all dischargers are not yet controlled by effective
measures.

The above is not to say that the regulatory side of the system,
which was at a dead end at the end of the 1960's, is disappearing as
pure economic theory would suggest. To the contrary, its evolution
and renovation have been noticeable over the last ten years, considering

both its budget and number of inspectors and the improvement of the
laws on which it is based. Clearly the regulatory side sets the
framework in which, on the whole, the basin agencies operate, even
though, on the fringes, coherence and coordination are not necessarily
guaranteed.

The regulatory system has kept its importance and has a bright
future for several reasons. First, some kind of licensing procedure
is necessary in any case, even if only to check on the discharge of
pollutants which are necessarily forbidden by law, e.g., carcinogens.
Second, experience has shown that industrialists are extremely
reluctant to modify, even marginally, their production technology,
even if the investment is paid back in one year, e.g., 100% return
on capital. Such behavior has been observed in many cases in the
Seine-Normandie Basin, for example, where an investment in a simple
cooling water recycling system--which would have been paid back in
less than two years--was not undertaken. There are individuals and
managers for whom economic incentive has no real meaning. Third,
moving from the ground of economics to that of the behavioral sciences,
the inducement for action is maximum when there is one stick, penalties
and sanctions for disregarding standards, from inspectors of the
Classified Establishments Service, and one carrot, technical advice,
loans, and grants, from a basin agency. It is even better when you
have two different actors to wield those different tools, so that the
roles are clear and the play can be more dynamic. An even greater
incentive for positive behavior can be achieved if those two actors
coordinate their actions behind the stage.

Thus, with the effluent charges system, complemented by designated
ambient water quality objectives to be achieved, the standards and
regulations regarding discharges achieved the equity, efficiency, and
political acceptability they needed to be effectively reached within a
few years by the vast majority of the dischargers.

Given the above, how can one assess, finally, the interplay of
the regulatory incentives and the effluent charge incentives for
municipalities and individual private activities? For municipalities,
the incentive effect of the regulatory system is fairly low because
it is difficult, if not politically infeasible, for the administration
to impose sanctions, i.e., you cannot send a mayor to jail. This
means the economic incentive is crucial; clearly, the system of grants
of all kinds that reduce the capital costs of treatment plants to
the municipalities provides the economic incentive. Furthermore, the
superpremium paid for a given level of efficiency of treatment is also
relatively high for municipalities, which also tends to increase the
economic incentive for reducing discharges by treatment. In fact, the
system is set so that relatively little additional regulatory incentive
is needed to push municipalities to take action to reduce discharges.
Table 20 shows that if one neglects capital costs, which is essentially
the case for municipalities because of various subsidies, the ratio
of the level of the charge divided by the marginal cost of treatment
is about 0.6 for primary treatment, and about 0.35 for secondary
treatment.

For private entrepreneurs, the situation is quite different. For
them, the regulatory incentive usually has real weight. Fines are

actually levied, court actions are taken in many places, Service of
Classified Establishment inspections are made. All this reduces
noticeably the margin of maneuver for dischargers. Nevertheless,
this kind of incentive has been insufficient in the past, because
in the 1960's, for example, the ambient water quality of French
waters rapidly worsened. What was needed then was some economic
incentive, but also, and even perhaps more important than a strict
economic incentive, an independent interlocutor (partner) which could
give technical advice on what actions the private entrepreneur could
take and how to take them to achieve the discharge standards. The
fact that this interlocutor also grants 30% of capital costs and that
the action taken will reduce the effluent charge, is not a bad point,
naturally, but this is often not the crucial point. This is why
the system of grants, premiums, and effluent charges is designed so
that it gives only about 15 to 20% of the incentive power to the
effluent charge, based on the cost of reduction by means of conventional
primary or secondary treatment.[4] Nevertheless, the combined effect of
regulations and economic incentives[5] is that private entrepreneurs do
take actions to reduce discharges.

The challenges ahead for French water quality management are:

(1) to complete the effluent charges system;

(2) to implement the Ambient Water Quality Objectives
Policy; and

(3) to open new frontiers in relation to (a) new pollutants,
e.g., nitrates and phosphates, bacteria, micropollutants,
heat and radioactivity, (b) nonpoint sources, e.g., urban
and agricultural, and (c) sewage collection systems.

NOTES

[1]Sludge handling and disposal provide the interface with the national Solid Waste Agency established in 1977.

[2]There are three approaches to deciding on the amount of aid to grant. The first is to take as a basis the capital cost of the treatment plant which would have been needed to reduce the discharge by the same amount as the proposed internal measures. Of the 100% financing, 70% has to be paid back by the industrial activity. The second is to take, as the base, the investment cost of the measures needed to reduce generation, minus the benefits of those measures, e.g., fewer inputs, materials recovery, by-product production. (The benefits do not include reduction in effluent charges.) The third is used when it is not known exactly before installation what the actual impact of the proposed internal measures will be. The basin agency finances 100% of the capital costs, plus fringe operating costs stemming from the experimental nature of the measures. If the measures perform as hoped, the activity pays back to the basin agency, with a not too high rate of interest, all costs which it would normally have had to pay in relation to such an investment (the capital costs, but not the fringe costs). If the measures do not work, that is, either a reduction in generation is not achieved or the cost of the reduction turns out to be economically infeasible, the activity does not have to pay back anything.

[3]As noted previously, the Ministry of Economics and Finance has never accepted this strange tax scheme which it doesn't control, and would have used conflict as an opportunity to scuttle the charges system.

[4]The percentage is the level of the charge divided by the marginal cost of treatment x 100%.

[5]It should be remembered that, in addition to paying effluent charges, private enterprises have access to loans at less than market rates of interest and are permitted rapid amortization of investments in antipollution equipment. The latter is an hidden subsidy, never accounted for in the computations of antipollution costs.

PART III

WATER QUALITY MANAGEMENT IN THE RUHR AREA

OF THE FEDERAL REPUBLIC OF GERMANY, WITH

SPECIAL EMPHASIS ON CHARGE SYSTEMS

Jochen Kühner and Blair T. Bower

PART III

Table of Contents

214

Chapter 12

INTRODUCTION

Part III describes water quality management in the state of Nordrhein-Westfalen in the Federal Republic of Germany. It emphasizes those economic, financial, institutional, and legal aspects of water management in the well-known river basin associations (Genossenschaften) Ruhrverband and Ruhrtal-sperrenverein, which aspects either have not yet been reported, or have been inadequately presented in the past. Because considerable literature exists concerning the Genossenschaften, including Kneese and Bower,[1] Irwin,[2] and Johnson and Brown,[3] the general background of those agencies—including history and purposes—does not have to be repeated in much detail. The following are the main aspects discussed:

- the administrative structure of the state (Land) for water quality management, and various discharge arrangements;

- effluent standards, state support for financing water quality management facilities, and the interaction of the Genossenschaften and the state's administrative structure for water management, including permitting of industrial direct and indirect discharges, and municipal discharges; and

- computation of the charges at both ends, i.e., for water intake and for discharges, for the Ruhrtalsperrenverein and the Ruhrverband, based on the budget and the physical and economic assessment procedures of the associations.

It should be emphasized that the report is based on the situation as of 1978. Although the following laws and their amendments will

influence the situation described herein, they will not change the
main points, especially with respect to the rationale for and computation of charges: (1) the Fourth Amendment to the Federal Water Act
of 27 July 1957, ratified on 26 April 1976 (Wasserhaushaltsgesetz);
(2) National Effluent Charge System ratified on 13 September 1976,
to be effective on 1 January 1981 (Abwasserabgabengesetz); and
(3) State Water Act of Nordrhein-Westfalen, ratified on 4 July 1979
(Landeswassergesetz).

The discussion herein concentrates on the Ruhrverband and the
Ruhrtalsperrenverein. First, the basic governmental and administrative
structure of the State of Nordrhein-Westfalen is outlined. Second,
some background on administration for water quality management in
Nordrhein-Westfalen is provided. Finally, the procedures for computing
charges in the Ruhrverband and Ruhrtalsperrenverein are discussed.

THE STATE (LAND)
OF NORDRHEIN-WESTFALEN

The state of Nordrhein-Westfalen has long been Germany's center
for coal production and the steel manufacturing and chemical industries.
It contains 81% of Germany's coal mines, representing about 90%
of the total coal production of the Federal Republic of Germany, and
72%, 31%, and 30% of the production capacities of the steel, petroleum
refining, and chemical industries, respectively. At a population of
17.1 million, it is the most densely populated state in the Federal
Republic of Germany.

The industrial center of Nordrhein-Westfalen is the Ruhr area, the largest industrial agglomeration of Middle Europe. About 5.6 million people live in this area of 4600 square kilometers. For about ten years, the traditional economic structure of the Ruhr area--coal mining and steel production--has undergone a restructuring process as a consequence of demand changes. In the years of, and following the recession of, 1966-67, the structural and growth problems became obvious, and resulted in significant employment difficulties, with a subsequent loss of the relatively high income standard, as compared to the remainder of Germany. This economic restructuring process has also influenced water quality management in the area. Contaminants of industrial effluents have changed; notably, phenols from cokeries have essentially disappeared.

This area has become known in the field of water quality management because of its river basin management agencies, the Genossenschaften. These agencies include: the Ruhrverband and the Ruhrtalsperrenverein (hereafter referred to as the RV and RTV, respectively) which are responsible for the Ruhr River, the southern portion of the industrial area; the Emschergenossenschaft, which is responsible for the River Emscher, draining the middle of the area; and the Lippeverband, which is responsible for the River Lippe, the northern portion of the industrial area. The Emschergenossenschaft and the RV/RTV were founded and established by special law as public corporations prior to World War I, in 1904 and 1913, respectively, by the legislature of Prussia with the approval of the King of Prussia. The Lippeverband was established in 1926.

The 1976 status of domestic wastewater treatment in Nordrhein-
Westfalen is shown in Table 22. Slightly more than 50% of the population was connected to secondary wastewater treatment facilities,
implying that most large cities have secondary treatment. Only about
5% was discharging from sewers without any treatment. Of course,
this tabulation is no indication, per se, of how well the plants
are operating, and hence of what is actually being discharged.

Table 22.   Status of Wastewater Treatment in Nordrhein-
            Westfalen as of 1 January 1976

% of Inhabitants	Status of Wastewater Treatment
51.2	connected to 1,041 secondary wastewater treatment plants
0.6	connected to 13 land treatment operations
30.3	connected to 175 primary wastewater treatment plants
5.7	connected to sewers without any treatment
87.8	
12.2	not connected to any central sewer system
100.0%	

NOTES

[1]Kneese, A. V. and B. T. Bower, Managing Water Quality: Economics, Technology, Institutions (Baltimore, Johns Hopkins University Press for Resources for the Future, 1968).

[2]Irwin, W. A., "Charges on Effluents in the U. S. and Europe." Prepared for the Council on Law-Related Studies (Cambridge, Mass., September 1971).

[3]Johnson, R. W., and G. M. Brown, Cleaning Up Europe's Waters: Economics, Management and Policies (New York, Praeger Publishers, 1976).

Chapter 13

WATER QUALITY MANAGEMENT IN THE AREA
OF THE RUHRVERBAND/RUHRTALSPERRENVEREIN
IN THE STATE OF NORDRHEIN-WESTFALEN

DESCRIPTION OF EXECUTIVE AND ADMINISTRATIVE
RESPONSIBILITIES IN THE STATE OF NORDRHEIN-WESTFALEN

After World War II, the state of Nordrhein-Westfalen was founded. It was put together from two major provinces: the western province of Prussia, Westfalen, and the northern portion of what was formerly the Rheinland. It then became one of the ten states (Länder) of the Federal Republic of Germany.[1]

Introduction to the Governmental
System of Nordrhein-Westfalen

In the state of Nordrhein-Westfalen there are two levels of governments of general jurisdiction. These are the state (Land) government, and the local government of either a county (Kreis) or county-free city (Stadt). Major cities are independent units of government, i.e., county-free cities; they conduct their local affairs independently of the county in which they are located. Thus the state is subdivided into areas governed by either counties or county-free cities, neither of which is subordinate to the other.

Every four years, the members of the legislative bodies of the state--Landtag--and of the counties and county-free cities--Kreistag

and Stadtrat, respectively--are elected. Based on party majority,
the Ministerpräsident and his ministers (heads of ministries) are
approved by the Landtag as the state government. The members of the
county and county-free city governments, i.e., heads of the various
departments, are named by the county mayor, Kreisrat, and the city
mayor, Oberbürgermeister, respectively, and then approved by the
respective legislative bodies. Local government is given the principal
responsibility for its own affairs. State government is a coordinator
and overseer without much power to control the conduct of local
affairs by counties and county-free cities.

Outline of Executive and Administrative Responsibilities in Nordrhein-Westfalen

Each government of general jurisdiction has its own administrative units, i.e., state authorities and agencies, and county and
county-free city authorities and agencies. The responsibilities are
delegated over various levels. This will be discussed separately
for each government of general jurisdiction.

Administrative responsibilities are divided into those that rest
on state authority and have therefore to be executed by state administrative authorities and those that evolve from the conduct of local
affairs, or have been transferred to local self-administrative
responsibilities by the state. In recent years, the state has put
many administrative duties under the authority of counties and county-
free cities, in order to relieve the state, and to further decentra-
lization of government.

There exist three different levels of administrative authority
to execute the responsibilities of the state, as illustrated in
figure 15.

1. At the highest level of administrative authority are
the ministers of the state (Land) ministries, e.g., Ministry of
Nutrition, Agriculture, and Forestry.

2. The second level is made up of five Regierungspräsidien
which together cover the entire state. Each of the Regierungspräsidien
encompasses in its area all the categories of administrative responsibilities that are present in all ministries of the state government.

3. The third level is made up of state agencies that have
no broad, but only very specific administrative authority and/or
technical advisory functions. They are located in each Regierungspräsidium, covering the region of the Regierungspräsidium, and are
overseen by the Regierungspräsidium in conducting their delegated
responsibilities. Their areal extension varies; it is, in general,
larger than the area of a county. There is overlap among them with
respect to areal jurisdiction, but every agency type covers the area
of each Regierungspräsidium.

The geographical distribution of agencies indicates that this
separation was done in order to execute effectively the authority
of the state within the whole state. Nordrhein-Westfalen is divided
into five Regierungspräsidien. In their respective areas they
administer all the responsibilities and authorities delegated to
them by the state level administration; the latter also oversees the
performance of the Regierungspräsidien. The head of each Regierung-

Figure 15.   Administrative Authorities of the State of
              Nordrhein-Westfalen

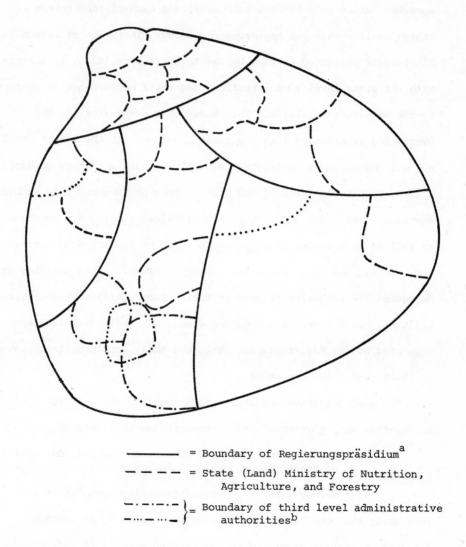

———————— = Boundary of Regierungspräsidium[a]

— — — — = State (Land) Ministry of Nutrition,
            Agriculture, and Forestry

—·—·—·— }
            = Boundary of third level administrative
—···—···— }   authorities[b]

[a]There are five Regierungspräsidien in the state of Nordrhein-Westfalen.

[b]There is a multitude of third level agencies within the area of each
Regierungspräsidium. Overlap exists among the agencies with respect to
areal jurisdiction, but the entire area of each Regierungspräsidium is
covered by each type of agency.

spräsidium, the Regierungspräsident, is named by the Ministerpräsident

of Nordrhein-Westfalen. Each Regierungspräsidium contains many

agencies having only limited authority; and each of them covers a

region smaller than the Regierungspräsidium. Their set of responsi-

bilities is delegated to them by the Regierungspräsidium, in agreement

with the state level administration, and their performance is overseen

by the Regierungspräsidium. This means that responsibility and

overseeing is arranged like a pyramid. However, at the lowest level

not all the existing administrative entities have authority to administer

all the responsibilities of the state. Some of them are only technical

agencies serving for their area in a technical supervisory function

as well as in a technical advisory function to the Regierungspräsidium,

for example, the water resources agencies, STAWA. Other agencies are

delegated the authority to conduct their state administrative responsi-

bilities, e.g., forestry and mining agencies. Their directors are

appointed by the Regierungspräsident, and their appointments approved

by their respective ministers.

The administrative structures of a county government and a

county-free city government are as represented in figure 16:

(a) County (Kreis)--county administration and administration
of county-incorporated cities; and

(b) County-free city--county-free city administration.

This means that the county administration--headed by an Oberkreis-

direktor, generally appointed by the elected council of the county

for a term of six to twelve years--also delegates some of its

responsibilities to adminstrations of county-incorporated cities

within the county. They are headed by Stadtdirektors. A county-

Figure 16.   Administrative Authorities of Counties and County-free Cities

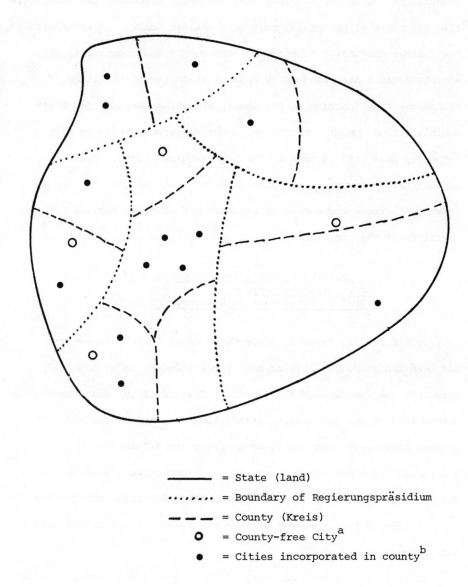

_____ = State (land)

•••••••• = Boundary of Regierungspräsidium

— — — = County (Kreis)

O   = County-free City[a]

•   = Cities incorporated in county[b]

[a] A county-free city has one administrative level.

[b] A county has two administrative levels: (1) the county itself; and
(2) incorporated cities within the county.

free city is headed by an Oberstadtdirektor appointed by the city's
council for twelve years. Thus this structure resembles the administrative structure of the state, only on a smaller scale. It is important
for further discussion to know that the county and county-free city
administrations also frequently conduct state responsibilities.
That means they function at the lowest executing level of the state
administration, under supervision of the Regierungspräsidium with
authority delegated to them by the Regierungspräsidium. This
administrative structure has been expanded over the last decades in
order to decrease the number of agencies and of civil service
positions at the state level.

Overview of Principal Executive and Administrative Responsibilities for Water Quality Management in Nordrhein-Westfalen

In the Federal Republic of Germany, water quality management
has been the responsibility of the states (Länder) under the 1957
framework law (Wasserhaushaltsgesetz). Changes of the Wasserhaushaltsgesetz in 1976 and subsequent introduction of a nationwide effluent
charge, Abwasserabgabengesetz, to be effective 1 January 1981,
and passage of a new state water law in Nordrhein-Westfalen in
July 1979 have altered the picture somewhat. But these changes are
largely ignored here, because they do not basically modify the information presented.

In most states, the authority of the state at the ministry level
with respect to implementation has been delegated to the various lower
level administrative authorities of the state. According to the above

general description of the administrative structure of the state, these authorities are distributed as described in the following paragraphs.

1. The Ministry for Nutrition, Agriculture, and Forestry has overall responsibility for water quality management in the state of Nordrhein-Westfalen. It interfaces with the federal government, and the other states; it oversees the implementation of responsibilities which it has delegated to other state administrative authorities; and it conducts all the affairs whose responsibilities it has not delegated.

2. The next state administrative level is the Regierungspräsidium. Some of the implementation authority of the Ministry for Nutrition, Agriculture, and Forestry is delegated to the state's five Regierungspräsidien. The respective departments of a Regierungspräsidium are called Dezernat (i.e., department), e.g., Dezernat for water and wastewater. They are overseen by the Ministry for Nutrition, Agriculture, and Forestry.

3. The lowest state administrative level for water quality management is represented by the administrations of counties and county-free cities. This means that each of the Regierungspräsidien has delegated some of its authority to the counties and county-free cities which exist in its area of jurisdiction, and oversees their performance. The existing state agencies at the lowest level, the STAWAs (Staatliche Ämter für Wasser und Abfall, i.e., state agencies for water and solid waste), have only technical advisory functions to the Regierungspräsidium and the city and county-free cities; they

have no administrative authority. Thus, each of the STAWAs provides necessary technical input to the Regierungspräsidium and to counties and county-free authorities within the Regierungspräsidium, for those water quality issues in the area for which each of the STAWAs is responsible. Generally, two to three STAWAs exist in the area of each Regierungspräsidium, covering the total area of each Regierungspräsidium.

At the state level, the Ministry of Nutrition, Agriculture, and Forestry has associated with it a staff agency, LAWA (Landesanstalt für Wasser und Abfall, i.e., state institute for water and solid waste), a technical agency without any administrative authority. This agency conducts difficult technical investigations for the ministry in preparation for the actions of the ministry. It also is assigned special tasks, such as monitoring the quality of the Rhine River within the jurisdiction of the state.

It is important to note that the distribution of administrative authority across the state administration differs from state to state in the Federal Republic of Germany. Therefore, the above structure cannot be applied to another state without caution. The state of Bavaria, in particular, is organized very differently.

BACKGROUND OF THE GENOSSENSCHAFTEN, RUHRVERBAND AND RUHRTALSPERRENVEREIN[2]

The Genossenschaften, which have been mentioned so far, were all established by special Prussian laws. Since 1937, however, a Federal law also exists which regulates establishment of water agencies

across Germany (Wasserverbandverordnung). The existence of this law
does not preclude establishment of agencies under a special law, if
special situations must be accommodated. This was demonstrated in
1958, when the Erft River Authority (Grosser Erftverband) was founded
in Nordrhein-Westfalen in order to resolve the water quality management problems associated with surface mining of lignite coal in the
area of Cologne. Thus, there are currently nine water resources
authorities in Germany, established under special laws: Emschergenossenschaft (1904), Ruhrverband (1913), Ruhrtalsperrenverein (1913),
Linksniederrheinische Entwässerungsgenossenschaft (1913), Lippeverband
(1926), Niesverband (1927), Schwarze-Elster-Verband (1928), Wuppever-
band (1930), and Grosser Erftverband (1958). All of them are located
in the former state of Prussia. The other approximately 15,000 water
and soil conservation agencies in Germany have the Federal Law of
1937 as their legal basis. This means that river basin authorities
could be created by special law in every state of Germany, if special
situations warranted. But only Prussia and, after the war, Nordrhein-
Westfalen, have established agencies by special law.

The Ruhrverband (RV) was created as a public corporation by
the Prussian legislature on 5 June 1913, by passing the Ruhr River
Pollution Control Law (Ruhrreinhaltegesetz), which was then signed
by the King of Prussia. On the same day, the Ruhr Reservoir Association, Ruhrtalsperrenverein (RTV), which has existed since 1899, was
also made a public corporation by the Prussian legislature by passing
the Ruhr Reservoir Law (Ruhrtalsperrenvereingesetz). Both agencies
were awarded the right to request mandatory contributions from their

members, and their jurisdictions were made equal to the watershed
of the Ruhr River.

The purpose of the Ruhrverband is to keep good water quality
in the Ruhr River and its tributaries.[3] Thus, the RV has "to build,
maintain, and operate the facilities that are required to prevent
pollution of the Ruhr and its tributaries by the individual members"
of the association. "The association is only obligated to undertake
water quality measures beyond those stipulated by Prussian Water
Laws when severe circumstances cannot be remedied in any other way..."[4]
The kind and number of required facilities, as well as their alterations
and expansions, are subject to the approval of the responsible minister.
In Nordrhein-Westfalen, this has been, since World War II, the
Minister of Nutrition, Agriculture, and Forestry. The minister can
delegate this authority to lower state water agencies covering the
Ruhr area, i.e., the Regierungspräsidien.[5]

The RV is also authorized "to build, maintain and operate, on
behalf of its members, facilities which are not required for achieving
the association's purpose but which are related to it. This applies
in particular to treatment plants which serve the special purposes
of individual members insofar as they exceed the purpose of the
association. The costs of such plants are to be borne by the party
which contracts for them."[6]

Members of the RV are:

1. the owners of mines and of other commercial or industrial
   enterprises, of railroads, and of other installations,
   that cause pollution of the Ruhr River system or that
   benefit from RV facilities;

2. the communities that lie partly or entirely within the
   drainage basin; and

3. the Ruhrtalsperrenverein (RTV)[7] on behalf of waterworks
   and other installations which withdraw water directly
   or indirectly from the river or its tributaries for
   other than energy generation purposes.[8] It should be
   emphasized that members of the RV include both direct
   and indirect dischargers.

Recently, military camps have been added to the membership of the RV
because of their pollution potential. The federal government objected
to contributing as "other installations;" it argued that wastewaters of
military camps were equivalent to those of a city and should be assessed
as part of a community. But in 1962 and again in 1966, the federal government lost two such cases in court.

Thus, besides the RTV, direct and indirect dischargers make up the
membership of the RV. Direct dischargers are municipalities and some
industrial activities; indirect dischargers are industrial, commercial,
and institutional establishments discharging into municipal sewer systems
which are connected to RV treatment facilities (as discussed below).

The purposes of the Ruhrtalsperrenverein are to replace the water that
has been withdrawn from the Ruhr River system to its detriment and to
improve the utilization of the hydropower potential of the river system.
A detriment to the river system is defined by the quantity of water which
is consumed, i.e., withdrawn and not discharged back into the Ruhr River,
in those time periods when the flow of the Ruhr River is less than
4.5 liters/second per 1 square kilometer of basin area.[9] Therefore,
the RTV's assignment is to: (1) build and operate its own reservoirs;
(2) support construction and operation of reservoirs not owned by the
RTV; (3) build and operate facilities to obtain water from the Rhine
River; and (4) build and operate other necessary facilities.[10]

Members of the RTV are:

1. waterworks, and other withdrawers of ground water and surface water from the Ruhr River and its tributaries, who withdraw more than 30,000 cubic meters per year; and

2. the users of the Ruhr River system's hydropower.

(Excluded are activities involving: [1] irrigation of meadows; [2] hydropower generation for self-supply of individual households and farms; and [3] hydropower generation of less than 10 horsepower.)[11] This implies that industrial and other activities can be members of both the RV and RTV, if they are dischargers and also self-suppliers of their drinking and/or process water. Otherwise, the individual memberships of the two agencies do not overlap.

In 1938, the staffs of the RV and RTV were merged in order to ensure better coordination, and a single directorate was created. However, the RV and RTV are still two separate legal entities, each having a separate board of directors.[12] Currently, a managing director and a technical director share the duties of the directorate.

Categories of members of the RV have been listed above. But industrial and commercial establishments only become members if their financial contributions exceed a threshold value. The by-laws of the RV currently stipulate that anyone whose contribution would amount to 1/100,000 of the year's total required contribution is a member. In 1970, this amounted to just over $100, and there were 1,229 members; in 1977, it amounted to $350, and there were 991 members (63 communities and 928 industrial, commercial, and other activities). The number of small industrial and commercial members has gone down

continuously because of the recession in, and other restructuring
forces of, the economy in the Ruhr area.

Irwin[13] reports that in 1970 there was some discussion within
the RV, in connection with recent amendments to the by-laws, about
increasing the fraction in order to raise the threshold of the contribution amount for becoming a member. This would reduce the administrative time and effort which the association's staff now spends with many
small members who, together, contribute only a very small portion of
the total contributions. The proposal was declined, however, on
the grounds that this very group, many of them basement electroplating
operations or their equivalent, could cause severe damage to the
basin's small brooks and streams if left outside the scope of RV
membership. Further, with a low minimum figure there was less chance
for business firms to feel unjustly treated because a slightly
smaller competitor fell below the line and did not have to make
payments to the RV.

In 1955, the administrative circuit court in Münster ruled that
membership may not be avoided on the ground that residuals are discharged
into a municipal sewer system, and not directly into the river system.
The same court handed down a decision on 1 January 1962, stating that
membership in the RV does not free members from their legal duty under
the state's water law to clean their own wastewaters, nor does it
permit them to require the RV to build the treatment facilities necessary
to do so.

The Community Tax Law of the state of Nordrhein-Westfalen
prohibits cities from levying sewer charges on an RV member for the

same benefits for which he has already contributed to the RV, thereby

giving RV charges precedence over local taxes. Thus, in theory,

establishments which are not RV members on the grounds that their

contributions are estimated to be below the threshold, should fall

under city regulations. The potential contributions of these non-

assessed establishments are added to the RV requested contribution

of the municipality in which they are located. Communities are

permitted to recover these additional charges from the non-RV

members according to their own formulae.

In general, municipal treatment plants are RV facilities. This

implies a sharing of responsibilities. Communities are responsible

for construction and operation of a sewer system to where a waste

treatment facility is located; then the responsibility of the RV

for wastewater treatment begins. Therefore, the RV has many options

from which to choose the regional scope and type of treatment system

according to an economic assessment. Exceptions to this sharing of

responsibilities might occur in isolated commercial and residential

developments. If a municipality cannot economically connect those

areas under its jurisdiction to its sewer system, and if the RV does

not agree to build a separate RV facility for those areas, the

municipality is obliged to build a separate facility on its own,

in order to meet effluent requirements established by the Regierungsprä-

sidium.

In most communities, stormwater is not separated from other waste-

water. Storm sewers are a municipal responsibility; overflow and

treatment facilities are the responsibility of the RV. This implies

the same sharing of responsibilities as in the wastewater case.

DETAILED DISCUSSION OF EXECUTIVE AND
ADMINISTRATIVE RESPONSIBILITIES FOR WATER
QUALITY MANAGEMENT IN NORDRHEIN-WESTFALEN

The overall distribution of responsibilities has been listed
above. In this section, some additional background on water quality
management and various other details will be added, with special
emphasis on the difference between areas outside and areas within
the jurisdiction of the RV/RTV.

Classification of Receiving Waters and
Effluent Requirements for Indirect and Direct Discharges

According to the state's water law (Landeswassergesetz, Section 2),
all surface waters--with the exception of wild waters and spa
springs--are classified into one of three categories, with first order
being the highest category. Groundwater is excluded from this
classification scheme. The assignment of a water body to a certain
category is made by the Ministry of Nutrition, Agriculture, and
Forestry on the grounds of the importance of the water resource in
relation to use. Some criteria for this categorization are flow,
shipping potential, water supply, and recreation. The stretch of the
Ruhr River from its confluence with the Rhine River to the city of
Witten, about 50 km upstream, was labeled first order; the remaining
part was second order. In the Landeswassergesetz of July 1979, only
two classes of surface waters are provided, and the stretch of
the Ruhr River classified as first order has been extended about 50 km
further upstream. The surface waters of first order are property of
the state unless they are the property of the federal government,

whose ownership is restricted to shipping canals. Surface waters
of second order are the property of the owners of the properties
abutting the water.

As will be discussed below, to discharge into the different
classes of surface waters requires a permit from the administrative
unit responsible for the particular stretch of the receiving water.
Requirements are imposed on the water quality of the discharge. Even
though the final decision has been at the discretion of the respective
administrative unit, the state in 1966 issued some guidelines on
discharge requirements for particular effluents. These guidelines
reflected the then state of "best practicable technology" as compiled
by a group of recognized technical experts from industries, universities, and governmental agencies. Residential effluents and industrial
effluents--such as those from meat factories, sugar beet processing
plants, pulp and paper mills, petroleum refineries, petrochemical
plants, and coal mines--were covered by these guidelines. The new
set of federal and state laws[14] passed in the period April 1976
through July 1979 has mandated that federal minimum requirements
(Mindestanforderungen) for effluent quality be developed for 57
activity categories.[15] A committee was formed for each of the categories,
with membership representing governments, industry, universities, and
consulting firms. In January 1979, the first standards under the new
legislation--minimum requirements according to current "best practicable technology"--were made final for residential effluents. It is
not known at this time to what extent in-plant modifications of
industrial facilities will be required, encouraged, and honored in
establishing requirements for industrial categories.

Financial Support by the State Government
for Water Quality Management

The state of Nordrhein-Westfalen provides financial support in
terms of grants to municipalities and water authorities, such as
the Genossenschaften, for the construction--but not for operation and
maintenance--of water quality management facilities. The following
types of facilities are supported: (1) mechanical-biological waste
water treatment; (2) stormwater treatment, including retention basins;
(3) wastewater and stormwater pumping works; (4) sewer systems; and
(5) pressurized pipelines.

For calculation of the maximum state support, a general calcu-
lative unit is introduced, the FE (Fördereinheit), which equals the
"Unit of State Support." For each of the above facility types,
different physical units are equated with one unit of state support,
i.e., 1 FE, as shown in table 23. Table 24 indicates the current
(1979) amounts of financial support--grants for construction--per
FE, provided by the state for various types and sizes of facilities.

Recipients of these state grants are the political/administrative
entities responsible for the construction of the facility. These might
be a city or a water authority.[16] It cannot be industrial activities
as they do not receive grants. Grants are administered by the state's
Ministry for Nutrition, Agriculture, and Forestry, and are made
available only after the facility has been approved by the respective
state administrative authority. (As of 1979, the reported time period
for approval was two to three years.)

Table 23. Physical Units Equivalent to One Unit of State Financial Support for Facility Construction, Nordrhein-Westfalen

Type of Facility	Physical Units Equivalent to One FE
Mechanical-biological treatment, including sludge treatment	1 PE, where a PE is defined as 60 grams BOD_5 per capita per day
Stormwater treatment, i.e., detention basin capacity	1 m^3
Wastewater and stormwater pumping works, i.e., pump capacity	1 liter per second
Sewer systems, including manholes and other appurtenances	1 meter length, for any given category of diameter
Pressure pipes	1 meter length, for any given category of diameter

Table 24. State Financial Support for Construction/Installation Provided per Unit of State Support, FE, for Various Types of Water Quality Management Facilities, Nordrhein-Westfalen, as of 3 April 1975[a]

Type of Facility and Size	Amount of Financial Support
Mechanical-biological treatment	
≤ 500 FEs	350
501 - 2,000 FEs	180
2,001 - 5,000 FEs	130
5,001 - 10,000 FEs	90
10,001 - 50,000 FEs	65
> 50,000 FEs	60
Stormwater treatment (capacity of basins)	
basins of reinforced concrete	100
basins with slight reinforce-ment of walls and bottom	50
basins, excavated without any reinforcement	15
Wastewater and stormwater pumping works	
≤ 50 FEs	900
51 - 100 FEs	650
101 - 200 FEs	500
201 - 500 FEs	350
501 - 1,000 FEs	300
1,001 - 2,000 FEs	250

Type of Facility and Size	Amount of Financial Support
Sewer systems, including manholes and other appurtenances--diameter (mm)	
250	100
300	106
350	112
400	118
450	124
500	130
. . .	
1,000	194
. . .	
1,500	274
. . .	
2,000	490
. . .	
2,500	702
. . .	
3,000	834
Pressure pipes	
≤ 150 mm diameter	25
200 mm diameter	35
300 mm diameter	45
400 mm diameter	55
500 mm diameter	70

[a] All units in Deutsche marks per FE (DM/FE). In June 1976, Units of State Support for stormwater treatment facilities and sewer systems were significantly increased. The increases ranged from 50% to 150%.

239

Currently (1979) there exist at least two additional funding mechanisms to provide capital costs of pollution control equipment or to provide some incentive to make such installations: the Lake Constance Program of the Federal Republic of Germany; and the low interest loans (Sondervermögen) of the European Recovery Program (Europdisches Wiederaufbauprogramm). In 1978, 300 million Deutsche marks of European Recovery Program money were assigned by the federal government to environmental protection. Of this total, 90 million deutsche marks were allocated to municipal wastewater handling by municipalities and water agencies. The European Recovery Program loans can be applied for through the respective state agencies. Commercial and industrial establishments can apply for these low interest pollution control loans, up to an amount of 300,000 Deutsche marks, directly at the German Bank for Reconstruction in Frankfurt am Main. In addition, Section 7d of income tax laws provide for tax deductions and short write-off periods for installation of pollution control facilities.

Water Quality Management in Areas
Outside and Inside of the Ruhrverband

There are certain differences in the way the administrative/legal tasks of water quality management are executed in areas outside and areas inside of the Ruhrverband. Figure 17 is presented to indicate alternative existing situations and to facilitate discussion. All parties shown in figure 17, both those inside and those outside the RV/RTV area, have to apply for some types of permission to withdraw water from and/or to discharge wastewater to the receiving water,

Figure 17.   Possible Configurations of Water Withdrawals and Wastewater
             Discharges in Relation to RV/RTV Area[a]

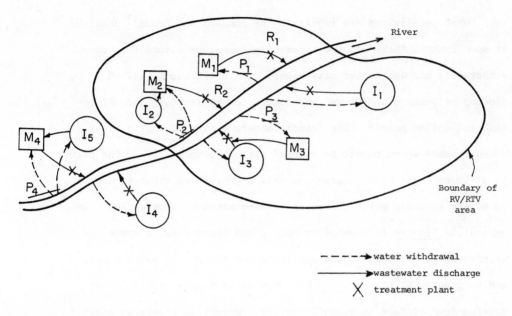

- - - - →water withdrawal

————————→wastewater discharge

✗   treatment plant

Legend

I₁: Industrial activity I_1 withdraws its own water, and discharges directly through its
      own treatment facility into the river.

M₁: Municipality is supplied by municipal water utility P_1 and discharges into RV
      facility. [b]

M₂/I₂: Municipality M_2 and industrial activity I_2 are supplied by private water utility P_2.
         Industrial activity I_2 discharges into the sewer system of Municipality M_2, which
         in turn discharges into RV treatment plant.

M₃/I₃: Municipality M_3 is supplied by private water utility P_3; Industrial activity I_3 is
         self-supplied. Municipality M_3 discharges into the treatment plant of industrial
         activity I_3.

M₄/I₅: Industrial activity I_5 discharges into the treatment plant of municipality M_4,
         located outside the RV/RTV area.

I₄: Industrial activity I_4 has its own water supply and discharge facilities outside the
      RV/RTV area.

M₄: Municipality M_4 is supplied by private water utility P_4, outside the RV/RTV area.

[a]All municipal treatment facilities in the Ruhrverband's area are RV facilities.

[b]See footnote "e" to table 25 for characterization of water utilities.

and they are also exposed to various charges. The loci of imposition
of permits and charges are shown in table 25.

What permissions are required? As Johnson and Brown[17] describe
at some length, there are two types of permissions issued for water
withdrawals and wastewater discharges: (1) Bewilligung, called
license by Johnson and Brown, and concession by Irwin;[18] and (2) Erlaubnis, called permit. The former confers a type of public right
(Wasserrecht) which cannot be revoked without compensation. The permit
is of lower legal status, granting only a short-term right to use water
in a very specific way. It can be revoked without compensation. In
general, a license is granted to public and private water works for
withdrawal of groundwater and surface water for public water supply
and to industrial plants with their own water supply facilities withdrawing from surface or groundwater. The permit is generally granted
for discharges of liquid residuals, even though a license for discharge
of "unpolluted" water is sometimes issued in conjunction with a license
for the withdrawal of water for, for example, cooling water. The
execution of the permitting process is somewhat different outside
of, than it is inside of, the RVT/RV area. Administrative fees have
to be paid by the party requesting a water withdrawal or wastewater
discharge, according to administrative regulations issued by the
Minister of Nutrition, Agriculture, and Forestry of Nordrhein-Westfalen.

Water Quality Management Outside the Ruhrverband Area

The responsibilities are distributed over the various state administrative levels as described in the following paragraphs.

Ministry of Nutrition, Agriculture, and Forestry. This ministry oversees all activities of lower agencies (implementation of laws and regulations), designs new laws, develops regulations, amends existing laws and regulations, distributes grant money for construction of water quality management facilities to municipalities, water agencies, and other eligible entities. (Industrial activities are not eligible.) It: channels and/or processes applications of municipalities for European Recovery Program loans; is the liaison to the Department of Interior of the federal government, which administers the federal water law; and cooperates with those ministries of other states that have been assigned the same responsibilities.

The Regierungspräsidium. The Regierungspräsidium establishes effluent requirements for individual direct dischargers and pretreatment requirements for indirect dischargers. It issues, changes, and withdraws discharge permits with all their stipulations, e.g., self-monitoring requirements, for first order receiving waters and records them in the Wasserbuch, a legal document. It issues, changes and withdraws water withdrawal licenses for all surface and groundwater waters and records them in the Wasserbuch, including all public and private withdrawals including those of agricultural operations; it levies administrative fees for licenses and permits. It oversees monitoring and inspection of dischargers, as conducted by STAWA or other staff agencies and institutes; for example, in certain parts of

Table 25. Water Withdrawal and Wastewater Discharge Permissions Needed and Agencies Levying Charges, RV/RTV Areas and Non RV/RTV Areas

Activity	Permission Needed[a]		Agencies Levying Charges	
	Water Withdrawals	Liquid Discharges	Water Withdrawals	Liquid Discharges
Industrial Activity I_1	x		RTV	RV
Industrial Activity I_2		(into public sewer system)		RV[b]
Industrial Activity I_3	x	x[c]	RTV	RV
*Industrial Activity I_4	x	x		M_4[d]
*Industrial Activity I_5	x	x	RTV	RV
Municipality M_1	x[e]		RTV	RV
Municipality M_2				RV
Municipality M_3		(into I_3 system)		RV (indirect discharger)
*Municipality M_4		x		
Private Water Utility P_2	x		RTV	
Private Water Utility P_3	x		RTV	
*Private Water Utility P_4	x			
RV Facility R_1		x[f]		RV[g]
RV Facility R_2		x[f]		RV[g]

Footnotes to Table 25

a Water withdrawal licenses are only issued by the Regierungspräsidium. Discharge permits are given by the Regierungspräsidium in case of a first order river; by a county/county-free city agency for dischargers into second order receiving waters. Permissions for RV facilities are handled by a special procedure.

b If the industrial activity is not a member of the RV, i.e., because its size is below the threshold criterion for mandatory membership, it is at the discretion of the municipality to recover from the industrial activity the costs imposed on the municipality by the discharge of the activity.

c Treatment of municipal wastewater in an industrial plant very seldom occurs. But it occurs at least once; the wastewater of Hagen-Kable is treated in a large biological wastewater facility of the Feldmühle paper mill. The RV contributes to the operating costs, and helped finance the plant.

d It is at the discretion of the Municipality M_4 to levy a sewer charge on Industrial Activity I_5.

e This implies that the agency P_1 that withdraws the water is a city department. Generally, drinking water supply is handled by corporations, either publicly owned by the major city they are servicing, or owned by a mixture of private investors and those public entities to which they deliver water.

f No regular permit required, but discharge specifications are set by the Regierungspräsidium in a Planfeststellungsverfahren (see figure 18).

g The RV levies the charges on those members which discharge into the RV facility.

* Activities located outside of RV/RTV area; all other activities are within RV/RTV area, as shown in figure 17.

Nordrhein-Westfalen this task has been delegated to the Institute
of Hygiene, a largely state-funded institute. The Regierungspräsidium also goes to court in case of repeated violations of the effluent
requirements; it is authorized to levy pre-specified penalties,
based on a legal catalogue of misdemeanors to be penalized by fixed
fines.[19] Finally, the Regierungspräsidium oversees the performance of
lower level agencies.

County and County-free City Agencies. These agencies--being
delegated certain executive authorities of the state--grant, modify,
and withdraw effluent permits for second order receiving waters with
all the same stipulations as the permits which are handled by the
Regierungspräsidium for first order receiving waters. But they do not
issue withdrawal licenses; those are only handled by the Regierungspräsidium. However, these agencies can also go to court and levy penalties
for misdemeanors, as can the Regierungspräsidium.

Water Quality Management Inside the Ruhrverband Area

The responsibilities are distributed as described in the following
paragraphs.

Ministry of Nutrition, Agriculture, and Forestry. This ministry
has all the same responsibilities described in the preceding section,
including distribution of grants to water agencies and Genossenschaften.
It oversees and controls the Ruhrverband and the Ruhrtalsperrenverein
to ensure that they operate within the limits of their laws, by-laws,
and regulations.

The Regierungspräsidium. The Regierungspräsidium has the same
responsibilities described in the preceding section with one major

exception. If RV facilities are reviewed for construction, the Regierung-spräsidium does not issue a permit to the RV with specific effluent requirements to be included in the Wasserbuch, but rather reviews, discusses, and agrees to effluent specifications for RV facilities in a formal hearing, termed the Planfeststellungsverfahren. The proposed specifications are publicly announced prior to the hearings in order to solicit public input. Afterwards, the final specifications are also made public. The status of the RV is delineated in the water law (Landeswassergesetz) of Nordrhein-Westfalen. Section 133 (2) of the water law states:[20]

> The Emschergenossenschaft, the Ruhrverband, the Ruhrtalsperrenverein. . . do not need a permit or license for their use of the respective receiving water. Plans made by these agencies have to be identified (ascertained) in a formal (public) procedure (hearing), if the governing water administration (Regierungspräsident) decides that the well-being of the affected population might be significantly affected or that disputes about the plans have to be expected or if the agency requests such a procedure.
>
> . . .the decision of the Regierungspräsident with respect to the plan is not preempted by the formal hearing (Planfeststellungsverfahren).

The Regierungspräsidium also orders monitoring of RV facilities to be done by the Institute of Hygiene, to which it has delegated the necessary power; and agrees to set pretreatment specifications for industrial facilities discharging into municipal sewers and then into RV facilities, i.e., pretreatment specifications for industrial indirect dischargers.

County and County-free City Agencies. These agencies have no delegated authority in the Ruhr Basin. All of the open hearings are conducted by the Regierungspräsidium.

Figure 18 indicates the basic differences between the administrative conduct of the Regierungspräsidium for RV and non RV facilities inside the Ruhrverband areas. The RV facilities are permitted through a formal hearing process; their effluent specifications are not put into the Wasserbuch in the form of a permit. This, however, does not imply that the specifications imposed via the formal procedure of an open hearing are not legally binding. What it means is that this procedure reflects the special position of the associations. e.g., Genossenschaften--granted under special laws--but does not imply a special function. The new state water law (Landeswassergesetz) of July 1979 no longer contains any special reference to the Genossen-schaften, including their exemptions from permits. To what extent administrative implementation has changed is not known.

It should be pointed out here, that other state agencies--or departments of the Regierungspräsidium--also put forward their demands for compliance with their regulations. For example, the land use agency has to judge how a treatment facility fits into the overall environment. The agency for commercial and industrial inspection (Gewerbeaufsichtsamt) judges the noise and air pollution impact against the existing effluent standards.

Monitoring Requirements, and Performances, Inspections and Evaluations

The administrative units of the state that issue licenses and permits impose monitoring and reporting requirements on both private dischargers and municipalities. Monitoring is performed by a technical

Figure 18.   Basic Differences between RV Facilities and
non RV Facilities within the RV Area under
State Water Law prior to July 1979[a]

Industrial Activity

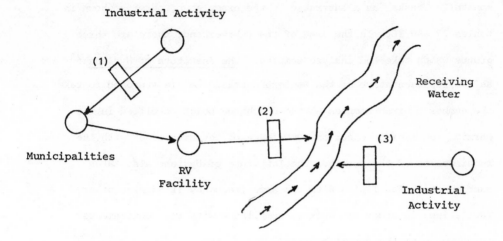

Receiving
Water

Municipalities

RV
Facility

Industrial
Activity

(1)   RV establishes pretreatment standards for the specific industrial
activity; the Regierungspräsidium imposes/legalizes these standards.

(2)   No discharge permit has to be issued for an RV plant. In general,
all necessary permissions to be given by the Regierungspräsidium
and other agencies--such as effluent conditions--are reviewed,
discussed, and agreed on in a Planfeststellungsverfahren (plan
setting procedure), in accord with section 133 of the state water
law prior to July 1979. The established effluent conditions are
not recorded in the Wasserbuch.

(3)   All discharge permits (Erlaubnis) issued by the Regierungspräsidium
are put into the Wasserbuch, in accord with section 45 of the state
water law prior to July 1979.[b]

[a]There do not exist explicit river quality standards but only use
specifications. However, river water quality standards will be set
under the new state water law of July 1979.

[b]Feedlots are included in activities requiring discharge permits.

agency or institute that has been delegated the power to execute this task. In recent years monitoring is also done by a discharger to whom a permit has been issued. The permit requests certain self-monitoring tasks, as illustrated in the examples of permits shown in tables 27 and 28. In the area of the Ruhrverband, there are three groups which take and analyze samples. The Institute of Hygiene,[21] as the representative of the Regierungspräsidium, is supposed to take the number of grab samples at the discharge point specified in the permit, in order to check on compliance of the discharger with the requirements of the permit. The Institute of Hygiene also takes samples at RV facilities with the same frequency it samples other facilities, in order to check on compliance with the requirements agreed upon in the formal hearing.

The RV takes samples of discharges for various reasons, two in particular. One is that the RV wants to identify discharge quality, in order to calculate its user charges as well as the performance of treatment facilities, in order to adjust the plant specific damage unit coefficients which are discussed in the next chapter. The other is that, in conjunction with samples of the water quality in the Ruhr River, it wants to set priority areas for future investments. It should also be noted that the RV monitors its own facilities. Whether or not the frequency of monitoring is higher than the one specified in the formal hearings is not known.

A third group, STAWA, takes samples of the receiving water as well as of effluents. Sampling of the latter has generally occurred only in relation to consideration of effluent permits. Therefore only samples

of the former two agencies were found in reviewing records of industrial
effluents. For a metal finishing plant, whose discharge permit is
shown in table 28, the parameters sampled by the Insitute of Hygiene
and the the RV are summarized in table 26. In general, one grab
sample is taken at the monitoring point at each sampling, which
generally is done four times a year. Based on discussion of the goals
of the two agencies and on table 26, there does not appear to be much
overlap with regard to their sampling. The Institute of Hygiene
samples the quality parameters specified in the permit and others
included in the effluent guidelines (Normalanforderungen), while
the RV concentrates largely on those parameters on which it has based
its plant-specific "damage-unit" coefficient. Apparently there does
not exist much coordination between the agencies with respect to
parameters to be sampled, and methods applied. Nevertheless, the
samples of the RV could not replace the samples of the Institute of
Hygiene because the state envisages conflict of interest because of the
membership of each discharger in the RV.

It is not obvious from investigation that formal criteria exist
for deciding on whether or not a discharger is in compliance with his
permit conditions. In general, a discharger is assumed to be in
compliance if four out of five samples are within limits with respect
to all water quality indicators.

According to the permits, personnel of the state agencies have
access to wastewater generating facilities and to the wastewater treatment facilities at all times. That means, appropriate inspections
can be done whenever they seem to be needed, as in cases of spills and

Table 26. Measurements Made by the RV and the Institute of Hygiene
at Comparable Locations in a Metal Finishing Plant

Measurement	Ruhrverband[a]	Institute of Hygiene[b]
Day/time of day	x	x
Temperature of water and air, °C		x
Transparency, cm		x
Color/turbidity	x	x
Material settled at bottom of effluent, yes or no		x
Odor		x
pH	x	x
Conductivity, $ohm^{-1}cm^{-1}$ at 20 °C	x	x
$KMnO_4$, mg/l[c]		x
NO_3^-, mg/l		x
NO_2^-, mg/l		x
NH_4^+, mg/l		x
Cl^-, mg/l		x
SO_4^{--}, mg/l		x
Hardness (total carbonate hardness)		x
O_2		x
Total Fe and dissolved Fe, mg/l	x	x
Total Mn, mg/l		x
Settleable solids, cm^3/l after 2 hours	x	x
Cu, mg/l	x	
Ni, mg/l	x	
Zn, mg/l	x	
Cr, III + IV, mg/l	x	
Cr, VI, mg/l	x	
General weather condition		x

[a]Grab sample of influent and grab sample of effluent from settling basin after neutralization

[b]Grab sample of effluent from settling basin and of discharge to river

[c]Milligrams per liter

suspicion of illegal discharges.

Discussions about sanctions for noncompliance, including noncompliance
of RV facilities, have always boiled down to the statement that the state
agencies have the power to go to court, or to levy pre-specified penalties,
or both. But no single case could be identified where such actions could
be clearly traced. In general, the cases cited were complicated by other
legal matters, so that violation of effluent requirements was only one
of many aspects.

Examples of Actual Withdrawal License and Discharge Permit

The foregoing sections have identified the agencies responsible
for licensing and permitting, and the general nature of licenses and
permits. Tables 27 and 28 contain examples of a license for water
withdrawal and of a permit for direct wastewater discharge, respectively.
Both of the plants are located in the RV area. The first is a metal
finishing plant which is self-supplying its intake water. The withdrawal
license--issued by the Regierungspräsidium--implies also a discharge
license for those waters which are not polluted and are only slightly
warmed up. However, although the license for the first is valid for
thirty years, the license for the latter is valid only for ten years.
For the withdrawal as well as the discharge, average flow figures
have been provided for four different time units; not all of them are
linearly related. The only self-monitoring requirement is a water
meter for the intake. The fee for the withdrawal license is calculated
on the basis of fee regulations drawn up by the Ministry of Nutrition,
Agriculture, and Forestry. The scale is generally very regressive; the

Table 27.   Example of a License for Water Withdrawal for a Metal
             Finishing Plant

Based on section 8 of the Wasserhaushaltsgesetz, and in connection
with section 22 of the Landeswassergesetz, the following license was
issued after review of the application and plan.

1. In addition to the conditions in the license of 1954, to withdraw
    groundwater as follows:

    up to 28 l/sec;   or up to 2,000 m^3/day;

    or up to 100 m^3/hour   or up to 600,000 m^3/year (implying
       (cubic meters per hour);   300 days at 2,000 m^3/day).

    The following additional amount can be withdrawn from the same source
    up to:

    18 l/sec;   1,300 m^3/day;
    65 m^3/hour;   400,000 m^3/year.

    The total amount can be used as process water and cooling water and
    can be partially consumed.

2. The license to discharge nonpolluted and only slightly warmed-up
    process water and cooling water up to:

    18 l/sec;   or 1,300 m^3/day;
    or 65 m^3/hour;   or 400,000 m^3/year,

    into the creek L., is also augmented such that the totals do not
    exceed:

    36 l/sec;   2,000 m^3/day;
    130 m^3/hour;   800,000 m^3/year.

3. The license for withdrawing groundwater will expire in 1995, i.e.,
    after thirty years and the one for discharge in 1975, i.e., after
    ten years. The following requirements are to be met:

    ● A water meter has to be installed at an appropriate point
      in the pumping work; the meter has to be read monthly and
      has to be checked for reliability every five years; all of
      the above have to be documented in the diary of operation.

Table 27 continued

- No cross-connection between public and the industrial water supply are permitted.

- The discharge of processing and cooling waters must be uncontaminated and should not contain any oils and fats; the temperature of such discharge should be 25°C at a maximum.

- Any changes have to be announced and have to be agreed on by the Regierungspräsident.

- The fee for the withdrawal license is 0.2% of a number calculated from a formula by the Ministry of Nutrition, Agriculture, and Forestry. This amount represents the administrative costs of processing the application for the withdrawal license.

1965
The Regierungspräsident

Table 28.   Example of a Permit for Direct Wastewater Discharge of a
             Metal Finishing Plant

Revokable Permit for discharge of sewage from sanitary facilities and
the metal finishing facilities up to an amount

$$\begin{array}{ll} \text{of} & 0.028 \text{ m}^3/\text{sec,} \\ \text{or} & 101 \text{ m}^3/\text{hour,} \\ \text{or} & 2,400 \text{ m}^3/\text{day,} \\ \text{or} & 630,000 \text{ m}^3/\text{year} \end{array}$$

after biological treatment and respective neutralization, through a pipe
of 550 mm under the following conditions:

1. a. runoff from precipitation, and groundwater, to be discharged
        directly into the receiving water;

    b. sanitary wastewater   ⎰ to be discharged
    c. effluent from neutralization ⎱ after treatment;

2. a. discharge must not contain any fats and oils, nor any
        dissolved and suspended metallic compounds;

    b. range of pH: 8.0 - 9.0;

    c. settleable solids, after 2 hours, $\leq 0.2 \text{ cm}^3/\text{l.}$

3. Temperature of effluent $\leq 25°\text{C.}$

4. Guidance has to be developed for the operation of treatment
    facilities. The personnel responsible for the maintenance
    must be named to the water administration. A diary of the
    operations has to be kept continuously.

5. Sludge from the neutralization and the sanitary waste treatment
    operation must be disposed of in a manner such that neither
    groundwater nor surface waters are contaminated by respective
    runoff or leakage. The disposal area has to be agreed to by
    the water administration.

6. The effluent from neutralization has to be self-monitored by
    continuously working monitoring and control equipment. Moni-
    toring results have to be included in the diary of operations.

7. At the points of discharge samples have to be taken four times
    per year by the state Institute of Hygiene in Gelsenkirchen.
    A manhole has to be constructed to enable samples to be taken.
    Results of analysis of samples must be reported, in the form of
    a letter, to the Regierungspräsident and to the county's public
    health administration.

Table 28 continued

8. Request for sampling four times each year has to be submitted
   to the Institute of Hygiene at once after receipt of the
   discharge permit. Respective costs have to be paid by the
   operator of the plant.

9. The discharge point has to be secured by a self-closing weir
   in case of high receiving water.

10. No cross-connection should exist between the public drinking
    water supply and the plant's own industrial water supply.

11. This permit is limited to those features contained in the plan
    that has been approved.

12. Alterations or expansions of the operations which would cause
    generation of additional and/or different wastewaters require
    a new permit.

13. Personnel of the Regierungspräsidium have to be permitted access
    to all operations at all times, as far as they are touched upon
    in the permit. On request, all the approved plans, the permit,
    and the diary of operations have to be made available at once.
    The water administration has the right to survey the operation
    of the plant at least once a year or, in case of violations,
    more often in accord with its own assessment. The plant is
    obliged to cover the costs of these surveys.

14. The permit is issued and can be revoked if more stringent
    requirements are needed. If other permits become necessary to
    prevent any other negative impacts, e.g., due to odor, they
    have to be obeyed.

15. The legal situation, set forth by civil laws, is not touched
    upon by this permit. The owner and/or operator of the plant is
    not relieved of the obligation to apply for other permits.
    The owner is also liable for all damage which is proven to have
    occurred due to the wastewater treatment facilities.

16. All regulations issued by other state or county/city agencies
    for the prevention of accidents have to be followed.

1963
The Regierungspräsident

more withdrawn per year, the lower the unit fee (Deutsche marks per
cubic meter). This schedule is used solely for the calculation of the
administrative fee associated with issuance of the license.

The permit for the metal finishing plant was issued by the Regierungspräsidium in 1963, after taking about one year to process the permit
application. The permit contains specific requirements for the
flow and quality of the effluent, for monitoring procedures, and for
inspections. But it is obvious that specifications with regard to
effluent quality are not very specific but are rather general;
only pH, settleable solids, oils and fats, metallic compounds, and
temperature are mentioned at all. Even though Fe is the effluent
parameter on which the RV's effluent charges for the plant are based,
Fe does not seem to be covered in the category of metallic compounds.[22]
Since 1963, the renewed permit has not yet been changed with respect
to detail. But that does not prevent the Institute of Hygiene from
analyzing the grab samples for parameters such as $KMnO_4$, NO_3, and
NH_4, not included in the permit. Such analyses provide a basis for
negotiations for the new permit, and a common ground for any actions
to be agreed to mutually in case of deficient effluents.

With this background on the administrative structure for water
quality management in the RV/RTV area, the discussion now turns to the
calculation of RV and RTV charges.

NOTES

[1] If West Berlin is included as a Land, the Federal Republic of Germany has eleven states.

[2] This section draws partially on the corresponding discussion of the Ruhrtalsperrenverein by W. A. Irwin in his paper "Charges on Effluents in the U. S. and Europe." This paper was prepared for the Council on Law-Related Studies, Cambridge, Massachusetts, September 1971.

[3] Ruhrreinhaltegesetz, section 1, subsection 1. The word "rein" translates as clean.

[4] Ruhrreinhaltegesetz, section 2, subsection 1.

[5] Ruhrreinhaltegesetz, section 2, subsection 3.

[6] Ruhrreinhaltegesetz, section 3.

[7] The membership of the Rurtalsperrenverin is discussed below.

[8] Ruhrreinhaltegesetz, section 4.

[9] Ruhrtalsperrengesetz section 2, subsection 2.

[10] Ruhrtalsperrengesetz, section 2, subsection 1.

[11] Ruhrtalsperrengesetz, section 1, subsections 1 and 2.

[12] See Kneese, A. V. and B. T. Bower, Managing Water Quality: Economics, Technology, Institutions (Baltimore, Johns Hopkins University Press for Resources for the Future, 1968).

[13] W. A. Irwin, "Charges on Effluents in the U. S. and Europe."

[14] The laws are: the Fourth Amendment to the Federal Water Act, 26 April 1976; the National Effluent Charge System, 13 September 1976, to be effective on 1 January 1981; and the State Water Act of Nordrhein-Westfalen, 4 July 1979.

[15] Wasserhaushaltsgesetz, section 7A.

[16]As noted above, there exists a multitude of water and soil conservation agencies in addition to the Genossenschaften.

[17]Johnson R. W., and G. M. Brown, Cleaning Up Europe's Waters: Economics, Management and Policies (New York, Praeger, 1976), p. 112.

[18]W. A. Irwin, "Charges on Effluents in the U. S. and Europe."

[19]Details of this catalogue and of its use were not available.

[20]This is the formulation in the state water law prior to the new law of 4 July 1979.

[21]The Institute of Hygiene (Hygiene Institut des Ruhrgebiets) is a state-founded and supported institute which is involved in many environmental monitoring tasks. It is a so-called "central institution," i.e., it is well supplied with state funds and has to be responsive to the needs of many agencies.

[22]Review of monitoring data for the plant reveals that sometimes Ni and Cr (III and IV) are detected in the effluent.

Chapter 14

COMPUTATION OF RUHRVERBAND
AND RUHRTALSPERRENVEREIN CHARGES

Johnson and Brown write:

> ...the actual charge system [of the Ruhrverband] departs
> from the textbook ideal. The basic charge does not vary
> with the location of firms; it is not a charge directly
> on pollution but is based rather on output or employees.
> Most strikingly, the pollution coefficient is assumed
> to be constant across all firms in an industrial group-
> ing unless there is proof to the contrary.[1]

Prior to discussing details of the Ruhrverband (RV) charge system which

should put the above quotation from Johnson and Brown in perspective, a

few sections of German laws will be mentioned, because they refer to the

core of each charge system, namely: (1) the manner in which costs are

distributed by associations to their members; and (2) the yardstick or

reference measure which is used to levy charges on individual members.

With regard to the first, sections 81 and 82 of the 1937 federal law

concerning establishment of water associations (Wasserverbandverordnung)

deal with the distribution of costs of all associations to their

members. The distribution is generally to be done according to

advantages derived from being a member of the association. Various

cost distribution schemes are considered feasible. For example, the

Genossenschaften employ at least three different schemes: (1) all the

facility costs are distributed to the members independent of which

member uses which facility; (2) each facility's cost is shared by the facility's users and the total membership, i.e., the specific users carry a certain percentage of the costs of the facility, such as 40% in the Lippeverband, while the remaining costs are borne by the membership as a whole; and (3) each facility's costs are borne only by its users, as in the Linksniederrheinische Entwässerungsgenossenschaft.

With regard to the second, the reference measure, the problem of most associations is the large number of members, e.g., about 1,000 in the RV, and their diversity, i.e., multitude of industrial and commercial activities resulting in wastewaters of very different characteristics. These factors lead to a dilemma caused by two desirable, but not necessarily coincident, characteristics of a charge: on one hand it should differentiate among members, but on the other hand it should keep the administrative burden and associated costs as low as possible. Section 86 of the 1937 federal water law (Wasserverbandverordnung) addresses this situation by "permitting approximate computations of charges." Similarly, section 6 of the Communal Tax Law of Nordrhein-Westfalen recommends a yardstick which is as close as possible to reality, but permits "approximating the charge" if it is uneconomic and administratively infeasible to derive the "real" charge as long as reality is not distorted. This has led to a situation in the large associations where members which are presumed to cause similar pollution or similar damage are grouped into one group. The waste loads of the members of each group are then measured at a number of plants, averaged, and, finally related to an easily measurable parameter.

After this brief reference to a few relevant sections of German

laws, this chapter addresses the following questions related to the

RV: (1) how is the pollution potential to which a charge is applied

assessed; (2) what is the basis for this assessment; and (3) how are

the budgetary requirements of the RV transformed into unit charges

and, ultimately, into charges to each discharger and withdrawer.

Similar questions will be covered with respect to the Ruhrtalsperren-

verein (RTV).

SOME BACKGROUND
ON RV CHARGES

The direct and indirect dischargers into the Ruhr River system

are required to make annual contributions to cover most[2] of the expenses

of the association, such as operating its treatment plants, paying

salaries and administrative costs, and paying the interest on capital

borrowed for investment in new treatment plants and related facilities.

The charges assessed are designed to obtain revenues to cover costs

incurred by the association.

A population equivalent (PE) of 54 grams BOD_5 per day[3] is the

yardstick against which waste discharges from industrial, commercial,

and residential activities are measured, including those containing

toxic substances. As described below, the PEs are converted to a

Bewertungseinheit (BE), i.e., an assessment unit. Freely translated,

it will be called a (physical) "damage unit" (BE). The total number

of each member's damage units is then multiplied by the unit charge,

Deutsche marks per damage unit--derived annually--to compute each member's

annual contribution.

The first step is the derivation of the "physical damage" associated with the discharge of each member. There are three basic groups of contributing members in the RV: (1) the Ruhrtalsperrenverein; (2) groups of commercial and industrial activities; and (3) municipalities. The RTV is legally obliged to pay 45% of the total fiscal contributions to the RV's budget. (Prior to 1935, the contribution was only 1/3, but was raised on the grounds of claims of damage done to the Ruhr River system by the transfer of water out of the basin by RTV members.)

The charge rates imposed on direct and indirect industrial and commercial dischargers are generally composed of: (a) a part for employee-caused pollution; and (b) a part for production-related pollution, i.e., industry/plant specific coefficients. For indirect dischargers there is also a part (c) for the amount of water discharged to an RV facility directly or indirectly via a sewer system. These three parts are translated into damage unit (BE) coefficients based on analyses of the wastewaters produced by a given activity or activity category.

(a) Every employee has a damage unit (BE) coefficient of 1/2. To simplify estimation of the number of employees, the number of working hours paid per year is used. Thus, 2,300 working hours equals one employee.

(b) The production-related population equivalent coefficients are based on the grouping to which a member belongs. If a member does not belong to any of the groups identified by the RV, the coefficient of the member is determined by special laboratory tests.

(c) The volume of wastewater discharged to RV facilities has a damage unit coefficient (BE) of 0.01 per cubic meter. The basis for determining the total amount discharged is the amount of water supplied. If the amount is estimated and the estimates appear doubtful, the RV may determine the amount discharged on the basis of its experience with other activities.

Contributions of municipalities are based on the municipal populations officially recorded by the statistical office of Nordrhein-Westfalen. The figure is adjusted for size of population and handling of wastewaters of non RV members, as described in the appendix to this chapter. Inhabitants who are not connected to a sewer system draining directly into the Ruhr or its tributaries are excluded from the assessment. Note that population is the basis for the charge on each municipality, not the actual discharge into the RV facility. This practice is analogous to the French procedure.

The most interesting group with regard to assessment is the second group, the industrial and commercial activities. Each member of this group has a coefficient characteristic of its group. There are 21 groups, some with several subgroups, including: coal mines and cokeries; metalworking operations; paper plants; textile plants; dairies; breweries; slaughterhouses; sauerkraut factories; hospitals; and gas stations. The coefficient specific to each group or subgroup is derived in a series of steps, based on bioassay tests described by Bucksteg.[4]

The procedure for determining the specific production-related coefficient for a group involves the following steps.

1. The effluent of the activity is tested against a residential effluent of known quality according to a standard bio-assay test, i.e., each of the effluents is added to a standard solution.[5] No indication is given of how the sample of effluent is obtained, e.g., single grab sample, composite of multiple grab samples, 24-hour proportional sampling every 15 minutes, etc.

2. Test results are plotted in a diagram as mg BOD_5 per liter added to the standard solution and ml (milliliters) of effluent to be tested per liter added to the standard solution, against the differences in $KMnO_4$ use (mg/l) between each of the solutions and the standard solution. This results in two curves: curve A reflects the results for the residential effluent; curve B for the effluent being tested.

3. A ratio is then derived from the two curves which expresses the number of liters of the effluent being tested which equals one population equivalent. This is the production-related coefficient.

4. Various plants of the same group are tested in this manner. The results are averaged and used to express the population equivalents of a group's wastewater, i.e., average number of population equivalents per cubic meter of effluent.

5. This series of tests is done for all industrial and commercial groups.

6. Provision is made to adjust these average pollution coefficients for each plant (activity) whose wastewater characteristics differ significantly from the average.

7. In order to ease the annual reporting of its members, the RV has related total annual population equivalents to total annual amount of activity, e.g., units of production, which can be easily reported by the members of each group. This results in a damage unit (population equivalent) per unit of activity. Examples are:

Cattle and pig slaughterhouses:	0.40 BE/1,000 kg of live weight
Sauerkraut:	0.36 BE/ton of raw cabbage
Milk storage tanks, without retention of cleaning water:	0.01 BE/100 hectoliters of milk
Milk storage tanks, with retention of cleaning water:	0.001 BE/100 hectoliters of milk
Yogurt production:	0.44 BE/hectoliter of yogurt
Metal finishing	0.31 BE/ton sulfuric acid used

Table 29 illustrates the calculation for a sauerkraut factory.

Table 29. Calculation of Number of Damage Units for a Sauerkraut Factory for Given Year

Employee-hours	230,000
Equivalent number of employees, i.e., employee-hours divided by 2300	100
Raw cabbage processed, tons	340,000
Wastewater discharge, to municipal sewer and ultimately into RV facility, m^3	780,000

Damage units for given year

$$= (100)(0.5) + (340,000)(0.36) + (780,000)(0.01)$$

$$= 50 + 112,400 + 7,800 = 120,250$$

Table 29 shows that the employees are counted each for 1/2 BE and each cubic meter of wastewater for 0.01 BE--a standard procedure for every plant--while the coefficient of 0.36 BE per ton of raw cabbage processed characterizes this particular industrial group. Tons of raw cabbage used is the easily reported amount which serves as a measure of the level of output of the activity.

Although the coefficients do not directly reflect production processes and wastewater characteristics, utilizing the coefficient instead of the value derived annually from all the samples taken of an activity's discharges, is the traditional approach of the RV. It must be assumed that the RV has identified those plant specific parameters that correlate well with the pollutants that are of most concern to the RV. Thus, as long as the production process of a member is very much in line with what the RV has labeled a standard process of a particular group, the member is assumed to have correspondingly standard wastewater and is assessed according to this coefficient. However, recent progress in laboratory methods and overall logistics permits taking and analyzing the many samples needed for the derivation of annual sample-based coefficients. (Previously, once established, coefficients were only adjusted in case of plant changes or disputes.) This would permit a change of the RV's assessment procedure, to reflect more accurately variations among individual activities.

Water use of industrial and commercial users is derived from an annual water use questionnaire. All the dischargers in the basin receive forms upon which they must record how much water they used for what purposes during the year. The physical unit which relates to

a unit charge is a weighted cubic meter of water withdrawn. The
weights are based on different use classes, which, in turn, are
differently set by the RV (and RTV) in computing the RTV members'
contributions to the RV and the RTV budgets, respectively, as is
illustrated in table 35.

DETAILED DISCUSSION
OF RV CHARGE SYSTEM

This section discusses all the basic pieces that lead to the
annual charges and the contributions of members.

The RV Annual Budget

The RV's income and expenditure statements and balance sheets
of capital investments for the years 1976 and 1977, respectively,
are shown in tables 30 and 31. These indicate the expenditures
incurred, and incomes and grants received. The regular contributions
of the RV and RTV members make up more than 80% of the annual income.
Including special contributions, about 90% of income is received
from the members of the RV and RTV. Other miscellaneous income
usually represents about 10%.

The largest expenditure item, operation, maintenance, and sludge
removal (of and from the RV facilities, excluding salary costs), is
about 28% of the total expenditures. Salaries and contributions to
social security represent the next largest item, about 20%, and wages
for non RV labor force, repayment of principal on bank loans, and
interest on loans, are of similar size, each of them between 13 and 17%.

Table 30.   RV Income and Expenditure Statement for 1976

Income	DM 10^9	%	Expenditures	DM 10^9	%
Rent and other contractural income	1.8	2.3	Salaries and contributions to social secutity	16.3	20.5
Interest and principal on loans made	0.8	0.9	General expenditures	1.1	1.4
			Wages	12.6	15.4
Contributions of RV members	69.7	84.9			
Special contributions of RV members	5.5	6.7	Operation, maintenance, and sludge removal/disposal	22.6	28.5
			Contribution to RTV for Bigge Reservoir	2.7	3.4
Various incomes	1.7	2.1			
			Interest on loans	10.1	12.7
Surplus from previous year	2.5	3.1	Principal of loans	13.5	17.1
Drawn from reserves	--	--	Special expenditures	0.8	1.0
	82.1	100.0		79.3	100.0

Balance Sheet of 1976 Capital Investments					
Assets	DM 10^9		Liabilities	DM 10^9	%
Treatment works and one administrative building	43.4		Grants from State of Nordrhein-Westfalen	16.5	38.0
			ERP loans	0.4	0.9
			Other income sources	0.4	1.0
			Bank loans	26.1	60.1
				43.4	100.0

Table 31.   RV Income and Expenditure Statement for 1977

Income	DM 10^9	%	Expenditures	DM 10^9	%
Rent and other contractual income	1.9	2.1	Salaries and contributions to social security	17.4	19.3
Interest and principal on loans made	0.8	0.9	General expenditures	1.2	1.4
Contributions of RV members	75.4	83.3	Wages	13.2	14.6
Special contributions of RV members	6.7	7.4	Operation, maintenance, sludge removal/disposal	23.7	26.3
Various incomes	1.4	1.6	Contribution to RTV for Bigge Reservoir	2.7	3.0
Surplus from previous year	2.1	2.3	Interest on loans	11.8	13.1
Drawn from reserves	2.2	2.4	Principal of loans	19.0	21.2
			Special expenditures	1.0	1.1
	90.5	100.0		90.0	100.0

	Balance Sheet of 1977 Capital Investments				
Assets	DM 10^9		Liabilities	DM 10^9	%
Treatment works and one administrative building	54.4		Grants from State of Nordrhein-Westfalen	22.5	41.4
			ERP loans	--	--
			Other income sources	2.8	5.0
			Bank loans	29.1	53.6
				54.4	100.0

Any excess income is transferred to next year's budget. The annual
capital investment balance sheets indicate that about 40% of the
necessary capital is provided via grants from the State, and about
60% has to be raised by bank loans--for which loans the annual repayment of principal and interest enters the expenditure statement.
The changes from 1976 to 1977 are minor, largely reflecting inflation.

Computation of Unit and Total Charges

The income and expenditure statements and balance sheets of
1976 and 1977 indicate the items in each year's budget. For the
purpose of the discussion in this section, it is assumed that the
RV budget for 1977, and the respective unit charge, C_{RV} , and the
individual contributions from the members are to be derived. The
procedure is as follows:

- In November 1976, the budget is approved for fiscal
  year 1977, which runs from 1 January to 31 December 1977.
- The relevant figures--population, numbers of employees,
  water use and wastewater discharge--are based on the
  year 1975. The respective questionnaire for the year
  1975 had to be submitted by 31 January 1976.
- The anticipated expenditures are based on escalating
  the previous year's budget and the five-year investment
  plan which is reworked every year.[6]
- The necessary total contributions are computed by sub-
  tracting anticipated "other" income--which is usually
  about 10% of total income--from estimated expenditures.

- Of the total contributions, 45% is paid by the RTV members
  (since 1935), and 55% by the RV members. For example,
  in 1977, of the regular contributions of 75.4 x 10^9 DM,
  33.9 x 10^9 DM were contributed by RTV members, and
  41.5 x 10^9 DM by RV members. If the RV's special contribution to the RTV, the Bigge Reservoir contribution, is
  figured in, the percentages become about 43% and 57%,
  respectively.

How is each member's annual contribution computed? As discussed
above, the annual damage formula--to derive the number of BEs--consists
of two parts for industrial/commercial direct dischargers, and of
three parts for an indirect discharger whose wastewaters enter an RV
facility. These parts are:

1. number of employees multiplied by a damage coefficient
   of 0.5;

2. plant specific unit damage coefficient, BE/base unit,
   multiplied by the number of annual base units, e.g.,
   tons of production; and

3. cubic meters of wastewater discharged to RV facility
   multiplied by a damage coefficient of 0.01.

In the case of a direct discharger whose discharge contains all of
his wastewaters, the last term is dropped.

The damage units for all non-municipal (point) discharges are
summed to obtain the industrial-commercial subtotal. Then the damage
units for each municipality are assessed and summed to obtain the
municipal subtotal. The total damage units, i.e., total number of

BEs, for the Ruhr River system are the sum of the two subtotals.[7]
The monetary requirement apportioned to RV members, based on the budget
for the year, is divided by the total number of damage units to yield
the unit charge for the year, C_{RV}, in DM/BE. Multiplying this unit
charge by each member's damage units yields each member's annual
contribution. Table 32 shows the evolution of the RV unit charge,
C_{RV}, over the last decade. A rapid increase beyond the 3-5%
inflation rate is obvious.

Table 32. Evolution of RV Unit Charges, 1964-1977

Year	Unit Charge, DM/BE, Levied on RV Members by RV	
	Municipalities	Others
1964	3.78	3.28
1965	3.92	3.66
1966	3.93	4.29
1967	4.15	4.75
1968	4.37	5.76
1969	4.90	
1970	5.08	
1971	5.82	
1972	6.98	
1973	7.98	
1974	9.03	
1975	10.20	
1976	11.21	
1977	11.92	

This computational procedure indicates that the difference
between the assessment of direct and indirect dischargers is made in
the physical assessment of each member's damage, i.e., number of BEs.
A direct discharger does not influence the capacity of an RV facility,
and hence the amount of wastewater discharged does not affect the damage
calculation. The total contributions made by each group in 1977 are
summarized in table 33. The high percentage, i.e., 70%, of the
contributions of municipalities to the total contributions by
dischargers implies a correspondingly high municipal share of operating
costs of the RV facilities into which the municipalities discharge.
This has significant implications in terms of incentives to industrial
dischargers to improve their effluents.

DETAILED DISCUSSION OF
RTV CHARGE SYSTEM

This section is patterned after the discussion of the RV charge
system. However, it should be remembered that RTV members--comprised
of water supply enterprises and other withdrawers from surface and
ground waters, and of hydroelectric energy generators--are assessed
in two ways because they have to make two contributions. One is
a contribution to the RTV budget; and the other is a contribution to
the RV budget, covering 45% of the contributions to the RV budget.

The RTV Annual Budget

Table 34 shows the major items of the 1977 income and expenditure
statement for the RTV. On the income side, the contributions of RTV

Table 33. Total RV Charges Paid by Sector in 1977

Sector	Special Contribution 10^9 DM	Regular Contribution 10^9 DM	Total 10^9 DM
Coal mines and coking facilities	0.29	--	0.29
Steel	0.07	9.23	9.30
Textile	--	0.13	0.13
Pulp and paper	--	1.66	1.66
Leather	--	0.44	0.44
Dairy	--	0.18	0.18
Brewery	--	0.93	0.93
Other facilities	0.10	1.15	1.25
Slaughterhouses	--	0.09	0.09
Federal government	--	0.36	0.36
Municipalities	6.23	28.57	34.80
Total payments by dischargers	6.69	42.74	49.43
RTV payments	--	32.76	32.76
Totals	6.69	75.50	82.19

Table 34.   RTV Income and Expenditure Statement for 1977

Income	DM 10^9	%	Expenditures	DM 10^9	%
Rent and other contractual income	4.3	7.4	Salaries and contributions to social security	7.9	13.8
Interest and principal on loans made	2.0	3.5	General expenditures	0.8	1.5
			Wages	3.1	5.4
Contributions of RTV members	45.2	78.6	Operation, maintenance and sludge removal	3.6	6.4
Profit of economic enterprises	0.3	0.5	Contribution to RV	32.7	57.7
Various other income	3.2	5.5	Interest on loans	3.5	6.1
Surplus from previous year	2.5	4.5	Principal of loans	3.9	6.9
			Special expenditures	1.3	2.2
Drawn from reserves	-0-	-0-			
	57.5	100.0		56.8	100.0

Note

In 1977, capital investments in various facilities of RTV reservoirs totalled about 7.8×10^9 DM. Fifteen percent of this amount was in the form of grants from the state of Nordrhein-Westfalen, 85% from bank loans.[8]

members are of overriding importance. The proportion represented by

these contributions is about 10% less than the proportion represented

by member contributions to the RV budget. It is also of interest to

see that the contribution to the RV makes up 55% to 60% of the RTV's

total expenditures. The item, salaries and contributions to social

security, is the only other expenditure representing more than 10%.

Each of four other items accounts for 5-7% of the expenditures.

Computation of Unit and Total Charges

All RTV members withdrawing water are assessed against the amount

of water they have withdrawn in specifically defined use classes. The

nine hydropower plant owners are exempted from contributions. Their

payments to the RTV, accounted for under various incomes, are based on

the additional energy they can generate because of augmentation of

low flows by releases from the RTV reservoirs.[9]

The computation of the RTV charge for a given year, e.g., 1977--

independent of the individual industrial activity's possible simultaneous

membership in the RV--proceeds as follows. Based on the 1975 questionnaire

on water use for each activity, the total cubic meters of water withdrawn

are computed. The amount of each water use activity is weighted by

use of specific loss coefficients. Table 35 indicates the four classes

of water use, A, B, C1, and C2. Each class has a different loss

coefficient, with the loss coefficients varying according to whether

they relate to the RV budget or to the RTV budget. Thus, the weighted

amount of water withdrawn is different for the computation of each

member's share for the RTV budget and for the computation of each member's share for the RV budget. The values of the loss coefficients have evolved over the years; the rationale for the particular figures could not be identified.

Table 35. Weighting Coefficients Used in Computing RTV and RV
          Charges, in Relation to Type of Water Use

Class A:	Water which is transferred outside of the Ruhr river basin
	Loss coefficient, A: 110%, RTV; 100%, RV
Class B:	Water used for drinking water or similar purposes within the Ruhr valley
	Loss coefficient, B: 40%, RTV; 65%, RV
Class C1:	Water used as process water within the plant within the Ruhr valley
	Loss coefficient, C1: 15%, RTV; 20%, RV
Class C2:	Water used as cooling water within the plant within the Ruhr valley
	Loss coefficient, C2: 6% RTV; and 6% RV

The unit charge, DM per m^3 withdrawn, associated with the necessary contribution of the RTV to the RV budget, is computed by dividing the necessary contribution by the total amount of water withdrawn and weighted per use class according to the RV loss coefficients. The direct unit charge, DM per m^3 withdrawn, associated with the RTV budget, is computed by dividing the necessary contribution to the RTV budget by the total amount of water withdrawn and weighted per use class according to the RTV loss coefficients. Table 36 shows the evolution of these unit charges on water withdrawals. Currently the RV unit charge is about 2 1/2 that of

Table 36. Evolution of RTV and RV Unit Charges on Water Withdrawals by RTV Members

Year	Total Withdrawal 10^6 m^3 [a]	Unit Charge, 10^{-2} DM/weighted m^3 by RTV	by RV
1964	1,206	2.39	2.10
1965	1,191	2.39	2.28
1966	1,199	2.60	2.39
1967	1,137	2.71	2.59
1968	1,046	2.98	2.74
1969	1,015	2.75	2.85
1970	1,087	2.39	3.05
1971	1,037	2.65	3.41
1972	1,158	2.41	3.92
1973	1,261	2.73	4.52
1974	1,322	3.17	5.02
1975	1,296	3.04	5.81
1976	964	2.99	6.58
1977	967	2.67	6.80

[a]Refers to the amount of ground and surface water actually withdrawn in the previous year

the RTV. Each of the two unit charges is multiplied by the weighted
amount of water withdrawn by each member, to yield the two contributions to be made by each RTV member.

Example of Calculation
of Charges on RTV Member

To illustrate the procedure, table 37 contains the calculation
of the water withdrawal charges for the sauerkraut factory identified
in table 29. According to the previous year's questionnaire, 1975, water
was withdrawn in all four use classes. Weighting those withdrawals with
the respective RTV and RV loss coefficients yields the weighted withdrawals, columns 4 and 6, respectively. Given the 1977 RTV and RV unit
charges, the total RTV and RV charges to be paid by the sauerkraut
factory are computed. Even though the weighted total withdrawal is
only slightly larger in the RV assessment than in the RTV assessment,
the RV monetary contribution is about 2 1/2 times that of the RTV
contribution.

The total contributions of the sauerkraut factory in 1977 were:

- to RTV for direct RTV budget -- 4.6×10^3 DM;
- to RTV for RV budget -- 12.1×10^3 DM;
- to RV for RV budget -- $1,433.4 \times 10^3$ DM;

for a total of about $1,450 \times 10^3$ DM. Clearly, the charges imposed by
the RTV are insignificant in comparison to those imposed by the RV
directly. The results also mean that the greatest payoff to the individual
activity would result from any actions reducing the generation of
wastewaters.

Table 37.   Calculation of Charges Imposed on an RTV Member, 1977

(1)	(2)	(3)	(4)	(5)	(6)
Water Use Class	Withdrawal, 10^3 m^3 per year	RTV Coefficient, %	Weighted Withdrawal, RTV (2) x (3)/100 10^3 m^3/year	RV Coefficient, %	Weighted Withdrawal, RV (2) x (5)/100 10^3 m^3/year
A	80	110	88.0	100	80.0
B	10	40	4.0	65	6.5
C1	250	15	37.5	20	50.0
C2	700	6	42.0	6	42.0
	1,040		171.5[a]		178.5[b]

[a]RTV charge = $(171.5)(10^3)(0.0267$ DM/m$^3)$

$\cong 4.6 \times 10^3$ DM

[b]RTV charge for contribution to RV = $(178.5)(10^3)(0.068$ DM/m$^3)$

$\cong 12.1 \times 10^3$ DM

NOTES

[1]Johnson, R. W., and G. M. Brown, <u>Cleaning up Europe's Waters:
Economics, Management and Policies</u> (New York, Praeger, 1976), p. 131.

[2]About 10% of the 1976 and 1977 RV budgets was derived from
miscellaneous sources, as shown in tables 30 and 31.

[3]According to recent German literature, the average population
equivalent has increased to 60 grams BOD_5/day, which is the value used
in determining the state's financial support (see table 23). It is
not clear whether or not the RV is still using the old figure of
54 grams in its internal calculations.

[4]Bucksteg, W., <u>Verfahren zur Bestimmung des Einwohnergleichwertes
beliebiger Abwasser</u>, Ruhrverband, Essen, Mimeo, 1958.

[5]Bucksteg, W., <u>Verfahren zur Bestimmung</u>. The procedure is
described in English in A. V. Kneese and B. T. Bower, <u>Managing Water
Quality: Economics, Technology, Institutions</u> (Baltimore, Johns Hopkins
University Press for Resources for the Future, 1968) pp. 240-251.

[6]It takes, in any case, a few years to move from the drawing boards
through the approval process, in order to begin construction of a
treatment plant.

[7]Note that no provision is made for discharges from nonpoint sources.

[8]It is not known whether or not the loans were at a market rate
of interest or a subsidized rate.

[9]Ruhrtalsperrengesetz, section 17.

Appendix to Chapter 14[1]

ASSESSMENT OF MUNICIPALITIES

The basis for municipal contributions to the Ruhrverband (RV) is the population officially recorded by the Statistical Office of Nordrhein-Westfalen. Only those inhabitants whose wastewaters drain into the Ruhr River or its tributaries are considered. Additional or above average pollution, which leads to higher costs for the association, are to be taken into consideration.

Municipal contributions are based on the following formula:

$$E = P \left[1 + 0.1 \sqrt{\frac{O^*}{P}} \left(n - \frac{P}{O^*} \right) + a \right],$$

where, E = the total number of population equivalents (to be multiplied by the unit charge rate in any given year to obtain the total charge);

P = 1 times the number of a city's residents whose wastewaters drain to the Ruhr River basin;

O^* = the minimum optimal population for an economically efficient sewage treatment plant;[2]

n = number of transfer points operated by the RV, i.e., points where the city sewage system transfers wastewater to the Ruhrverband; and

a = the factor for: other treatment measures; above average pollution; and for the amount and quality of wastewaters of enterprises which do not reach the minimum amount of contribution provided in the bylaws for membership.

The "a" factor takes into consideration the association's expenditures for other installations and measures, as far as these are not regarded as equivalents to treatment plants or exceed their scope, and as far as these measures are not to be assessed as special ones for the benefit of a restricted number of members. In addition, other expenses, for example those caused by special pollution or via the discharge of above average amounts of water of unusual quality, are to be incorporated in the "a" factor.

The effect of the transfer point factor, $0.1\sqrt{\frac{O*}{P}}\left(n-\frac{P}{O*}\right)$, is to require an additional contribution if fewer than 30,000 people exist per transfer point in the municipality. However, a reduction of the contribution results if the RV operates with a smaller measure in relation to the number of inhabitants, or the population for each transfer point exceeds $O*$.

If a treatment plant or its equivalent is put into operation in the first half of the year, the "n" factor increases accordingly for the entire year. If it is not put into operation until the second half of the year, the "n" factor is not altered until the next year.

The coefficient $0.1\sqrt{\frac{O*}{P}}$ has the effect of assuring that the costs of a treatment plant constructed for a particular community are borne by it only within the bounds of the principle of distributing all costs in proportion to members' pollution loads.

If a community discharges wastewaters from treatment facilities not located in the Ruhr River basin into an RV treatment plant, it is to be taken into account by an addition to the number of inhabitants, P. The magnitude of the addition is based on the amount of wastewater discharged. For each cubic meter of wastewater, a P figure of 2.5 is employed.

To illustrate how this apparently formidable formula works, assume
a municipality of 30,000 people with one RV treatment plant. Within the
municipality there is a municipal slaughterhouse, which does not meet the
minimum requirements for RV membership. Its discharge is assessed at
4500 damage units (BEs). Based on this assessment, the "a" value in the
municipal assessment formula = 4500/30,000 = 0.15. Therefore,

$$E = 30{,}000 \left[1 + 0.1\sqrt{\frac{30{,}000}{30{,}000}} \left(1 - \frac{30{,}000}{30{,}000}\right) + 0.15\right]$$

$$E = 30{,}000 \ [1 + 0 + 0.15]$$

$$E = 34.5 \times 10^3 \ \text{BEs}$$

where 0* = 30,000

This figure is to be multiplied by the unit charge rate for the given
year. At the unit rate for 1977, 11.92 DM/BE, the contribution of the
municipality would be a little over four hundred thousand DM.

Municipalities may receive discounts, e.g., municipalities--or
counties--for which the RV has not been of service and which have less
than 5,000 people, or less than 5,000 in their two largest villages
together. Municipalities where less than half of the population is
connected to a central water supply may receive a 50% reduction in the
required contribution.

A municipality may also receive a discount, upon written application,
if the number of its residents having their own properly operated house
treatment facilities or septic tanks which meet legal standards exceeds
the average number significantly, since such a municipality contributes
to the pollution of the Ruhr to a lesser extent because of pretreatment,
or not at all if they do not discharge any wastewaters. The average
number of residents who pretreat or do not discharge is to be regarded

as fifteen percent. In figuring the municipality's percentage, the
number of pretreaters is to be reduced by 50%, the number of nondischargers
counted in full. A discount will be granted if the 15% average is
exceeded by at least 5% and the following conditions are fulfilled.

1. The wastes retained in private treatment facilities are
disposed of without damage and do not reach either an RV
treatment plant or a waterway.

2. The RV requires no treatment facility for the group of
pretreaters or nondischargers.

3. No discount has been previously granted under the above
provisions.

The excess over the 15% average is the basis for the discount. The
conditions regarding the timing of discount applications, testing, and
effectiveness of facilities apply as appropriate.

Contribution amounts determined for industrial and commercial
enterprises and municipalities on the basis of these procedures may be
changed to the detriment of the member if:

1. no water use report is filed and the ensuing estimate is
determined on the basis of newly available facts to be too low;

2. the information provided by a member in his report is
determined by the Board of Directors, without consideration
of his fault, to be incorrect and the member does not prove its
correctness upon demand;

3. the conditions for receiving an assessment discount, in
the judgement of the Board of Directors, no longer exist
and the member does not prove their existence upon demand; and

4. a greater extent of pollution has occurred as the result

of a change in business operations since the delivery of the

assessment report.

NOTES

[1]The material in this appendix is modified from Irwin, W. A.,
"Charges on Effluents in the U. S. and Europe " prepared for the Council
on Law-Related Studies (Cambridge, Mass., September 1971).

[2]This number is presumably adjusted periodically. The rationale by
which any number is derived is not known.

Chapter 15

SOME INSIGHTS INTO THE LIPPEVERBAND

A major difference between the Lippeverband and the Ruhrverband/
Rurtalsperrenverein is the fact that the water of the Lippe River is
not used for drinking water purposes. Therefore, the multiuse problems
encountered by the RV/RTV are not relevant to the Lippeverband. This
has led to a different management approach and a different charge system.
Another important difference is that the Lippeverband does not cover
the total basin of the Lippe River, but only the downstream part from
Lippborg on.

There are three types of problems with respect to which the
Lippeverband makes investments and carries out operation and maintenance
activities. These are: (1) flow in the Lippe River and its tributaries,
including handling of wastewater; (2) stormwater drainage; and (3) maintenance of the channels for the prevention of floods. Only the first
is discussed herein, concentrating on the basis and derivation of the
contributions of Lippeverband members in relation to their wastewater
discharges. The Lippeverband has four types of members:

    1. The waterworks that import drinking water into the

    Lippe River basin;

    2. Activities that withdraw directly from surface and/or

    groundwater of the Lippe River and its tributaries;

3. Generators of wastewater, subdivided into municipal,
commercial-industrial, and coal mines; and

4. Activities that withdraw water and also generate wastewater.

The following basic steps are performed in the computation of the
contribution of (charges assessed against) the members for wastewater
management:

1. preparation of the annual budget (income and expenditure)
of the Lippeverband, which budget is similar to that of the
RV, but in which the costs and revenues associated with each
treatment facility continue to be separately identified;

2. compilation of water withdrawals, Q_W;

3. assessment of the degree of water usage, by multiplying
the amount Q_W by a factor reflecting different uses of water
within the activity, i.e., Q_W x U;

4. assessment of the unit damage, Z, in the dilution-damage
system;

5. assessment of the total effluent, Q_E, for each wastewater
generator and for each of the Lippeverband wastewater treatment
plants;[1]

6. derivation of the "degree of damage" by multiplying the
unit damage, Z, by the effluent flow for each wastewater
generator and each treatment facility;

7. distribution of the expenditures estimated in the budget for
each wastewater treatment plant according to a fixed scheme
between water users, 11%, and wastewater generators, 89%, for
every wastewater treatment plant separately;[2]

8. the distribution of the 89% of the costs of each facility to be borne by wastewater generators is 40% to the users of each facility and 60% to the total membership;

9. based on the degree of damage, Q_E x Z, and the degree of water usage, Q_W x U, and the costs to be distributed among the members, the following are derived:

    a. a general Lippeverband unit charge for water withdrawal;

    b. a general Lippeverband unit charge for wastewater discharge, in relation to the portions of the costs of the various treatment works to be paid by the total membership; and

    c. the unit charge associated with each treatment plant in relation to the portion of the costs to be borne by the dischargers into the plant;

10. based on the various distributions of costs, the total contribution of each member is derived by summing his individual contributions.

Thus the contributions to be made to the budget are separated into total membership contributions and discharge facility specific contributions, and therefore different unit charges per degree of damage are derived. To make the procedure clear requires further explanation of how the degree of damage is determined, and an example.

Much has been written about the Lippeverband and especially its damage assessment formula. This formula is based on a dilution procedure, the result of which is a dilution unit (Verdünnungsfaktor).[3] What is frequently not reported, or is misrepresented (as by Johnson and Brown),[4] is the fact that this dilution unit has to be transformed into a damage unit,

Z (Schädlichkeit), which is then modified by the amount of wastewater discharge to yield the degree of damage associated with the discharge. The following assessment formula has been used since 1958:[5]

$$D = -1 + \frac{S}{S_p} + \frac{1}{2}\frac{B}{B_p} + \frac{1}{2}\frac{P-30}{P_p} + F,$$

where, D = dilution factor;

S = materials subject to sedimentation, cm^3/l;

S_p = permitted S in cm^3/l, (S_p = 0.4);

B = BOD_5 in mg/l after sedimentation;

B_p = permitted BOD_5 in mg/l, (B_p = 20);

P = potassium permanganate oxygen used in mg/l after sedimentation;

P_p = permitted potassium permanganate use in mg/l, (P_p = 21); and

F = toxicity to fish as determined by dilution method.

The resulting dilution, D, reflects the dilution water that is needed to reduce the concentration of contaminants to such a degree that the concentration is not detrimental to the water quality of the Lippe River.

The dilution, D, is transformed into damage units, Z, according to the data shown in table 38. The data in the table have evolved over time and reflect the Lippeverband's experience of how wastewaters affect water quality in the Lippe River.[6]

Next the water withdrawal, Q_W, and the water effluent, Q_E, must be derived. The former, when multiplied by the water use factor, U, yields the degree of water usage; the latter, when multiplied by the

Table 38.   Relationship between Dilution and Damage, Lippeverband

D, dilution	Z, damage units
0 - 4	1
4 - 8	2
8 - 12	3
12 - 16	4
16 - 20	5[a]
20 - 28	6[b]
28 - 36	7
> 36	theoretically up to ∞ [c]

[a] Domestic wastewater has been shown to have a dilution of
D = 18.5, corresponding to five damage units.

[b] Slope of curve changes.

[c] This continues theoretically until ∞, which is in contrast to
the Emschergenossenschaft where Z_{max} = 20. Beyond that level a
special assessment is conducted.

damage factor, Z, yields the degree of damage. The data for the
derivations are obtained each year by the Lippeverband via a questionnaire
which each member must complete and submit. Information requested
includes such items as:

<u>Production-Related Data</u>

average number of employees
number of operating days
level of activity, e.g., tons of input processed or output
    produced
size of plant area, m^2

<u>Water Data (unit = m^3)</u>

input of water:
    from water works   from own withdrawal
                                    surface water
                                    ground water
proven water losses
effluent of water as:
    cooling water   other

usage of total water input as:
    cooling water   processing water
    sanitation   other

effluent discharged to:
    municipal sewer system   Lippe River
    tributary   Lippe Canal

The questionnaire has changed only slightly over the years with respect
to the requested data. Not all the data are used directly in the
assessment process; some are used for checking consistency of information
and some are just needed to evaluate the plant's continuity.

The withdrawals from surface and/or ground water are weighted
differently depending on how the water is used. If the water is used
for purposes requiring a quality similar to drinking water, the factor

is 1. If the water is used as processing water and discharged without major consumptive losses, the factor is 0.38. If the water is used exclusively for cooling and is discharged without any chemical or biological contaminants and only slightly heated up, the factor is 0.06.[7]

In order to calculate the general unit charges and the specific discharge costs, the yearly costs estimated for each wastewater treatment plant have to be distributed among the water users and the wastewater generators. The following rules are applied.

- The waterworks that import drinking water into the Lippe basin have to pay 1% of the total annual contribution.

- Activities which withdraw from surface waters and/or groundwaters of the Lippe River and its tributaries have to pay 10% of the total annual contribution. Each withdrawer's charge equals the user's weighted withdrawal, i.e., degree of water usage, times the unit usage charge. The charge is computed by dividing the allocated portion of the total costs, i.e., 10%, by the aggregated degree of usage by the total membership.

- The remaining 89% of the yearly contributions are distributed to the wastewater generators; 40% of this sum to be contributed by the local wastewater generators discharging to the specific local facility, and the remaining 60% to be contributed by the membership as a whole.

The foregoing is illustrated in table 39, which shows: the estimated 1976 costs of the wastewater treatment plant of city H; and the allocation of those costs according to these rules.

Table 39.   Allocation of Estimated 1976 Costs of the Wastewater
Treatment Plant of Municipality H

	10^6 DM
(1) Estimated 1976 costs	2.76
Necessary Contributions	
(2) Water utility of city of H, importing water into Lippe basin, 1% of (1)	0.03
(3) All withdrawers of water in Lippeverband, 10% of (1)	0.28
(4) Estimated costs less 11%, (1) − [(2) + (3)]	2.45
(5) Contribution of total Lippe-verband membership, 60% of (4)	1.47
(6) Contribution of dischargers into wastewater treatment plant of city H, 40% of (4)	0.98

The aggregate contribution of the dischargers into the wastewater
treatment plant of city H is allocated among those dischargers according
to their respective degrees of damage.   Illustrative data are shown
in table 40.

Table 40.   Allocation of Estimated 1976 Costs of Wastewater Treatment Plant of City H to Direct Dischargers

Costs to be allocated		980×10^3 DM
Necessary Contributions	Damage Units, $Q_E \times Z, 10^6$	Charge, 10^3 DM
Residences of city	36.53	544
Brewery	3.78	56
Pipe rolling and finishing plant	0.47	7
All others	25.27	373
Totals	66.05	980

Note that these figures yield an unit charge of 14.8×10^3 DM per damage unit for the wastewater treatment plant of city H.

The contribution of the pipe rolling and finishing plant toward 1976 estimated costs is derived as follows. The plant discharged 93.3×10^3 m^3 into the wastewater treatment plant of city H in 1975. The analysis of the effluent showed a dilution of 19, yielding a value of the damage factor, Z, of 5, from table 38. The degree of damage of the plant then is $93.3 \times 10^3 \times 5$, or 466.5×10^3 damage units, rounded to 0.47×10^6 in table 40.

Although the derivation of charges in the Lippeverband differs somewhat from that in the Ruhrverband, the objective in both cases is the same, namely, to cover the annual costs incurred by the respective agencies. If the revenues in any given year are less than expenditures--because less water is used and/or less wastewater is discharged--the Genossenschaften draw upon reserves. Conversely, in years when revenues exceed expenditures, the excess is put into the reserve fund.

NOTES

[1]Steps 2, 3, and 4 are based on the yearly questionnaire submitted by each water user/wastewater generator to the Lippeverband. Step 5 is based on analysis of samples.

[2]Note that in the Lippeverband, the costs of each individual treatment plant are distributed separately among each plant's respective users and to the overall membership.

[3]See Kneese, A. V. and B. T. Bower, Managing Water Quality: Economics, Technology, Institutions (Baltimore, Johns Hopkins University Press for Resources for the Future, 1968), pp. 245-248.

[4]See Johnson, R. W. and G. M. Brown, Cleaning Up Europe's Waters: Economics, Management, and Policies (New York, Praeger, 1976), p. 124.

[5]The formula was developed at a time when coal and steel represented the main industries in the area, and when wastewater was discharged without any treatment, or perhaps, after some mechanical treatment.

[6]However, the wastewater characteristics have been gradually changing because of changes in the region's industry and changing production and treatment processes. This has led to a review of the assessment process. A modified formula and transformation to damage have been suggested in which: the parameter COD--used in the German national effluent charge--would replace $KMnO_4$; wastewater flow would become a parameter; and the result of the formula would be a loading factor, B, in population equivalents. This would mean that the transformation to damage could be internalized in the formula.

[7]No information is provided on how the weighting is accomplished when cascade water use, or even simple recycled water use, exists, e.g., the same cubic meter of water is used first for cooling, then for process water, then for plant cleanup.

Chapter 16

CONCLUSIONS: THE ROLE OF EFFLUENT CHARGES
IN THE GENOSSENSCHAFTEN AREAS

In the Genossenschaften areas, effluent standards and effluent charges
have existed side by side for decades. Effluent standards relating to
individual activities--both for discharges into municipal wastewater
treatment plants and for discharges directly into water bodies--are
established by one or another of the regular state (Land) agencies. The
extent of compliance with such standards and the types and extent of
sanctions imposed for noncompliance are not clear. No case of prosecu-
tion was found in the investigation. Responsibility for determining
compliance and imposing sanctions lies with the state agencies.

Effluent charges are imposed by the Genossenschaften--such as the
Ruhrverband, the Ruhrtalsperrenverein, and the Lippeverband--which are
associations established in Nordrhein-Westfalen under special laws. The
charges are imposed in order to cover the annual expenditures--operation and
maintenance, debt service, administration--incurred in performing activities
to meet the mandates of the associations. The procedures for allocating
costs and for assessing the imputed damages associated with any given
discharge have developed historically, making it difficult to find
rationales for everything.

In addition to effluent standards and effluent charges, other
implementation incentives exist. Outright grants to cover a portion of

the capital costs of municipal wastewater treatment facilities are made
by the state of Nordrhein-Westfalen. Loans for capital investment in
pollution control facilities are available to both municipalities and
private activities at less than market rates of interest. Tax laws
permit rapid depreciation for investments in pollution control facilities.

THE FUTURE

The national effluent charge system is to go into effect on
1 January 1981. Presumably the objective of the system is to induce
further reductions in discharges, beyond those achieved by the imposition
of effluent standards by the states (Länder), and/or with respect to
substances not currently included in discharge permits. Because the
national system involves some parameters which are not currently used
in monitoring discharges, relevant measuring and monitoring procedures
and definitions of noncompliance will have to be established, and new
laboratory capacities developed.

Table 41 shows the criteria to be used in the national system for
establishing the damage units associated with any given discharge.
The total number of damage units is multiplied by the unit damage
charge to yield the effluent charge for an activity. The unit charge
is set at 12 DM for 1981, to rise to 40 DM by 1986. These charges,
in addition to those levied by the Genossenschaften, may be sufficient
to induce some reductions in discharges beyond the levels specified in
discharge permits.

Table 41. Criteria to be Used for Assessment of Damage of Discharges, National Effluent Charge System of the Federal Republic of Germany

Criteria	Unit of measurement, quantity per year	Damage units per unit of measurement
Settleable substances for which organic content $\geq$ 10%[a]	1 m^3 settled	1.0
Settleable substances for which organic content $\leq$ 10%	1 m^3 settled	0.1
Oxidizable substances, as measured by COD[b]	100 kg	2.2
Hg and compounds[c]	100 g Hg	5.0
Cd and compounds[c]	100 g Cd	1.0
Toxicity toward fish	1,000 m^3 wastewater	0.3 G_F[d]

[a]Measurement procedure: reduce amount by 0.1 ml/l wastewater beforehand.

[b]Measurement procedure: reduce amount by 16 mg/l wastewater beforehand. Silver sulfate is the catalyst in the dichromate method specified.

[c]Measurement procedure for Hg and Cd: atomic absorption spectrometer.

[d]G_F is the dilution factor, e.g., down or up to nontoxicity. If wastewater is discharged in coastal waters, toxicity is not considered for those substances whose content is based on salts which are comparable to those in ocean water.

THE AUTHORS

Blair T. Bower is Consultant in Residence, Resources for the Future, Washington, D.C.

Remi Barre is the Director (President) of GERPA, a consulting firm in Paris, France, specializing in studies of environment and natural resource management. He is a native of Paris.

Jochen Kühner is a private consultant in Cambridge, Massachusetts. He is a native of Hamm, Westfalen, Federal Republic of Germany.

Clifford S. Russell is Associate Director, Quality of Environment Division, Resources for the Future, Washington, D.C.